21st Century COE Program

Slavic Eurasian Studies No. 11

Dependent on Oil and Gas: Russia's Integration into the World Economy

Edited by
Shinichiro Tabata

Slavic Research Center, Hokkaido University

First Published in 2006
Publisher: Osamu Ieda
Editor: Shinichiro Tabata

Dependent on oil and gas : Russia's integration into the world economy
/ edited by Shinichiro Tabata.
Sapporo : Slavic Research Center, Hokkaido University, 2006.
(Slavic Eurasian studies ; no. 11)
    p. ; 21cm.
ISBN 4–938637–40–5
1. Petroleum industry and trade -- Russia (Federation)
2. Gas industry -- Russia (Federation)
3. Capital movements -- Russia (Federation)
4. Corporate governance -- Russia (Federation)
I. Tabata, Shinichiro, 1957–.

Printed in Japan

Table of Contents

Preface

PREFACE

This volume represents joint work carried out by a team of Japanese economists in the field of Russian studies with the aim of understanding Russia's integration into the world economy. This joint project was financed by the fund of the Ministry of Education and Science in the form of a grant-in-aid for the comprehensive research on Russia's integration into the world economy from 2001 to 2004. Support for this project will continue through funding from the same Ministry as a grant-in-aid for the research project entitled, "Russian capitalism and the flow of financial resources" from 2005 to 2008, and also funded by the 21st Century COE Program "Making a discipline of Slavic Eurasian studies: meso-areas and globalization" from 2003 to 2007.

In Chapter 1 of this volume, we summarize the characteristics of our joint research and our major findings. In Chapter 2, we present findings estimating the real size of the oil and gas sector in the Russian economy. Actually, Chapter 2 is reprinted from *Eurasian Geography and Economics* (Vol. 46, No. 1, 2005). We would like to thank the editor of that journal for the permission to include that publication in this volume.

Chapters 3 and 4 showcase two of our favorite methods of research using input-output tables and balance of payments, respectively. Chapter 3 analyses the influences of the oil industry throughout the whole economy. Chapter 4 investigates the problem of capital flight from Russia. In the final chapter, Chapter 5, a micro analysis of the oil and gas industry is conducted.

Shinichiro Tabata

Sapporo

June 13, 2006

Chapter 1

Oil and Gas in the Economic Transformation of Russia

Shinichiro Tabata

INTRODUCTION

Our studies of Russia's integration into the world economy have been characterized, first of all, by deep and minute analyses of the official statistics of Russia. First we gathered statistical materials from official sources and carefully read up on the literature of Russia's statistical methodology. Largely owing to Masaaki Kuboniwa's initiative, we have been in direct contact with statisticians working at the Goskomstat Rossii (State Committee on Statistics of Russia) or the Rosstat (Federal State Statistical Service of Russia) since 2004, and also at the Interstate Statistical Committee of the CIS.

The post-1991 period of economic transformation also represents the period of transforming statistics in socialist countries from statistics based on notions of Marxist economics to those internationally regarded as standards. In the sphere of national income statistics, this means the transition from the Material Product System (MPS) to the System of National Accounts (SNA). Kuboniwa played the role of an adviser to the Goskomstat Rossii during this transition and many Russian statisticians were invited to Japan to learn the statistical methods of Japan. At the same time, researchers in our group took advantage of consultations and meetings with Russian statisticians in order to deepen our knowledge of the specific practices of Russian statistics.

Secondly, our research has featured the intensive use of input-output tables. On the one hand, we investigated in the 1980s input-output tables compiled by the Soviet Union. We have learned much from Vladimir Treml's studies on the reconstruction of Soviet input-output tables. By using these investigations, later we were able to deeply consider characteristics of the Russian economy and statistics, and its continuity from the Soviet period. On the other hand, in the 1990s Kuboniwa and Treml were appointed as foreign advisers to the Goskomstat Rossii for the compilation of Russian input-output tables based on the SNA. Owing to this, we have been able to obtain direct information on Russian input-output tables until now.

The third characteristic of our research has been the emphasis on foreign economic aspects of the Russian economy. Contrary to popular understandings that the Soviet economy had been little influenced by foreign trade, we paid close attention to the role and influence of exports of oil and imports of agricultural products in the Soviet era. In the period after 1991, when international factors played a larger role than before for Russia, we intensively investigated not only foreign trade statistics, but also the balance of payments statistics, which also began to be compiled in early 1990s. Akira Uegaki revealed important features of the Russian economy by comprehensively analyzing the balance of payments statistics (Uegaki, 1999). Uegaki also investigated regional aspects of foreign economic relations (Uegaki, 2001, 2002, 2003).

Fourthly and lastly, we have attached great importance to the state budget statistics. Although published state budget statistics are poor in content, the information included in them is indispensable for analysis of the Russian economy. Especially, as we are interested in flows of financial resources, the distribution of profits or value added through the state budget is critically important. Because in the budgetary system of Russia, regional budgets have occupied a significant part, we have also studied regional budgets and inter-budgetary transfers (Kuboniwa and Gavrilenkov, 1997; Tabata, 1998, 2003).[1]

In this chapter, we summarize our findings in recent joint research.

---

[1] In this connection we have visited many regions in Russia with the aim of investigating inter-budgetary relations in Russia. In the period from 1998 through 2003 we visited 15 regions ("subjects of the Federation" in Russian terminology), including Moscow city, Ivanovo and Voronezh Oblasts, Sankt Petersburg city, Vologda Oblast, Nenets AO (Autonomous Okrug), Rostov and Nizhnii Novgorod Oblasts, Tatarstan Republic, Sverdlovsk and Tyumen Oblasts, Khanty-Mansiisk AO, Novosibirsk and Irkutsk Oblasts and Khabarovsk Krai.

OIL AND GAS IN THE RUSSIAN ECONOMY

One of the most significant achievements in our study of the Russian economy is the finding concerning the real size of the Russian oil and gas industry. The official figure for the share of the oil and gas sector in Russian GDP can be derived from the input-output tables compiled by the Rosstat. The problem with the official figure is that it is very low, namely 6.5 percent in 2002. We argued that part of value added produced by oil and gas has been recorded in trade sectors as trade margins and in the transportation sector as transportation margins, and there were net taxes on oil and gas included in net taxes on products. This was caused by low prices on oil and gas extraction preferred by oil and gas companies in order to minimize tax payments (see Chapter 2 of this volume for a more detailed explanation).

Kuboniwa and the Rosstat jointly investigated this problem by using input-output tables and calculated "actual" contributions of the oil and gas sector to GDP (Kuboniwa, 2002, 2004; Kuboniwa et al., 2005). These data are summarized in Table 1, which shows that the share of value added produced by oil and gas in Russia's total GDP was 18.9 percent in 2002. The volume of value added recorded in the trade and intermediation enterprises has been the largest since 1999 among enterprises shown in Table 1.[2] This was due to the sharp increase in world oil prices, resulting in the increase in the share of total value added of the oil and gas sector (see Fig. 1).

As shown in Table 2, more than half of trade margins of oil and gas have been generated from export activities. This share was highest in 1999 and decreased to approximately 50 percent in 2001–2002. But, except for oil processing, this share was still high in 2001–2002. These data also suggest that the volume of value added recorded in the trade sector was influenced by exports of oil and gas, and accordingly by world oil prices.

This reminds us of continuity and discontinuity carried over from the Soviet period. The fact that economic growth depends heavily on exports of oil and gas has not been changed. In the Soviet period, however, all

---

[2] The decrease in value added of transport enterprises in 1999 was due to a change in statistical classification. In 1999 Gasprom's headquarters began to be classified as an economic unit in the foreign trade sector (Tabata, 2002, p. 612).

revenues earned by exports of oil and gas were included in the state budget as so-called "special foreign trade earnings," which was estimated to have accounted for 7–8 percent of the total state budget revenue (Tabata, 1996, p. 141). At present, the majority of these revenues seem to be captured by oil and gas companies. In this connection, we have investigated tax revenues of the state budget from oil and gas and revealed some improvements in tax collection from the oil and gas industry in recent years (Tabata, 2002, 2006a).

The price differences on oil and gas between domestic and world markets represent implicit subsidies for consumers of oil and gas in Russia, including manufacturing companies. This fact also constitutes "continuity" from the Soviet period. The present situation concerning implicit subsidies was discussed in Shiobara (2004) and Tabata (2006a).

As the next step in our study of the actual contribution of oil and gas to Russian GDP, we have analyzed Russian economic growth in real terms, explicitly taking into account the actual size of the oil and gas sector mentioned above (Tabata, 2006b).[3]

DUTCH DISEASE

As a result of our joint research, we argued that one of the most significant causes of the great depression of the 1990s in Russia was the high exchange rate of the ruble maintained by the competitive exports of fuels (Kuboniwa and Tabata, 1999; Tabata, 2000). By this high exchange rate, imports of consumer goods and manufactured goods were promoted, resulting in a reduction in the production of domestic manufacturing industries. This phenomenon is usually called "Dutch disease."[4]

Changes in real exchange rates of the ruble are shown in Fig. 2. Rapid appreciation in the period from 1992 through 1995 is outstanding. In the period from 1996 until mid-1998, when the so-called corridor sys-

---

[3] The same kind of analysis to measure the actual contributions of the oil and gas sector to GDP growth was also attempted in Ahrend (2006).

[4] See Chapter 3 of this volume for more detailed explanation about Dutch disease. Yasushi Nakamura analyzes the influence of oil and gas industries throughout the Russian economy by using social accounting and national accounting matrixes (Nakamura, 1999, Chapter 3 of this volume).

tem was applied, the high ruble rate was maintained.[5] This was the period when GDP continued to decline in Russia, except for a slight increase in 1997.

After the financial crisis in 1998, when the official rate of the ruble was devalued by half in real terms (see Fig. 2), the Russian economy recovered quickly. The effect of devaluation seemed to continue until 2001. In that period, manufacturing industries, including machinery, metallurgy, chemicals and textiles, recorded high rates of growth (Tabata, 2006b, pp. 102-103). Owing to sharp increases in oil prices which started in 2000, GDP growth accelerated and continued after 2001.

Due to increases in oil export revenues, the ruble has been gradually appreciated in real terms since 1999 (see Fig. 2). The real exchange rate of the ruble reached its pre-crisis level already in the beginning of 2004. Therefore, we observed some symptoms of Dutch disease, i.e., slowing in growth rates in manufacturing industries (ibid.). Then, we have to consider the reason why the present Russian economy is immune from Dutch disease. Tabata (2006a) argues that price differences in oil and gas between domestic and world markets that have widened in recent years partially explain this question. That paper also argues the considerable increases in state budget revenues, promoted by the simplification of taxation on oil and gas and the direct link of tax rates on oil to world market oil prices, contributed to continued economic growth in Russia.

Tabata (2006a) concludes that future perspectives for the Russian economy depend significantly on the use of the energy windfall revenues through the state budget and the Stabilization Fund. As pointed out in Gavrilenkov (2004) and others, the diversification of the economy is, undoubtedly, the best remedy for Dutch disease.

CAPITAL FLIGHT

For the diversification of the economy and for the renovation or innovation of manufacturing industries, foreign investment with high technology is definitely needed. Russia's particular economic situation can be characterized by its abundance of financial resources and its lack of do-

---

[5] See Chapter 4 of Uegaki (2005) for the exchange rate system and policy of Russia in this period.

mestic investment. Financial resources have been accumulated through a huge amount of trade surplus due to oil and gas exports. But the ratio of gross fixed capital formation to GDP has been low because of expected low return from investments in manufacturing industries suffering Dutch disease (Tabata, 1997). The difference between gross savings and gross capital formation (gross fixed capital formation plus changes in inventories) represents a surplus of current account plus statistical discrepancies, which is considered to be the upper ceiling of the possible level of capital flight. Tabata (1997) calculates these amounts using SNA statistics.

Uegaki tried to calculate more accurately the amount of capital flight by comprehensively analyzing the balance of payments statistics (see Uegaki, 1999, 2004a, Chapter 4 of this volume). He estimated the amount of capital flight as the sum of such items as "errors and omission," "foreign cash currency" and "unpaid export (import) charge."

Kuboniwa (2002, p. 2) suggested that one of the main sources of capital flight might be trade margins in relation to the oil and gas sector. This is another "role" of oil and gas in the economic transformation of Russia.

If we explicitly add government in the balance of savings and investment explained above, we could observe how private investments have been undermined not only by capital flight, but also by a significant amount of state budget deficits in the period 1995-1997 (Uegaki, 2004a, pp. 33-34).

POLICY ISSUES

Our statistical analyses of financial flows and our minute survey of changes in the economic system lead us to the evaluation of economic policies adopted by the Russian Government in the period of economic transformation. For example, we intensively analyzed the causes of the financial crisis in 1998 (Popov, 1999; Chapters 3 and 4 of Shiobara, 2004; Chapter 4 of Uegaki, 2005).

Uegaki (2004b) has written the most systematic evaluation or criticism of economic policies pursued by the Russian Government and recommended by the IMF and foreign advisers. Our consensus is that the financial crisis in 1998 implied the complete failure of policies recommended by the IMF, characterized by global liberalism, including complete liberalization of currency markets, adoption of the corridor system, issuing a great amount of short-term domestic bonds and liberalization of state bond markets for non-residents.

We also investigated the problems concerning Russia's accession to the WTO (Konno, 2002). Special attention is paid to issues related to the oil and gas industries, including the difference between domestic and world market prices on natural gas as criticized by the EU.

Russia's economic policies toward CIS countries were critically analyzed in Tabata and Suezawa (2004) and Tabata (2005). These studies reveal that economic relations with CIS countries still play an important role in the economic development of Russia, especially in the sphere of foreign trade, including oil and gas. The specific role of small-scale integrations, such as the Eurasian Economic Community, the Union State with Belarus and the Common Economic Space with Belarus, Kazakhstan and the Ukraine, has been taken into consideration.

CONCLUDING REMARKS

Since 2005 we have launched a new joint project, entitled "Russian capitalism and the flow of financial resources." The aim of this new project is to clarify the specific characteristics of Russian capitalism by analyzing the flow of financial resources. We will continue our statistical analyses on the flow of financial resources on the basis of national accounts, input-output tables, balance of payments, budget, customs and banking statistics. We plan to deepen our analysis by investigating the balance sheet of oil and gas companies. We will continue to pay special attention to the factors of foreign economic relations with both CIS and non-CIS countries.

REFERENCES

Ahrend, Rudiger, "Russia's Post-crisis Growth: Its Sources and Prospects for Continuation," *Europe-Asia Studies,* **58**, 1:1-24, 2006.

Gavrilenkov, Evgeny, "Growth in Russia and Economic Diversification," in Tabata and Iwashita (2004), 93-121, 2004

International Financial Statistics (IFS). Washington, DC: International Monetary Fund, annual and monthly.

Konno, Yugo, "Problems of Russia's Accession to the WTO," *SRC Occasional Papers*, No. 87, 1-22, 2002 (in Japanese).

Kratkosrochnye ekonomicheskie pokazateli Rossiiskoi Federatsii. Moscow, Russia: Rosstat, monthly.

Kuboniwa, Masaaki, "An Analysis of Singularities of Russia's Marketization Using Input-Output Tables," *The Journal of Econometric Study of Northeast Asia*, **4**, 1:1-13, 2002.

Kuboniwa, Masaaki, "Russia's Oil and Gas Reconsidered," in Tabata and Iwashita (2004), 143-159, 2004.

Kuboniwa, Masaaki, and Evgeny Gavrilenkov, *Development of Capitalism in Russia: The Second Challenge*. Tokyo, Japan: Maruzen, 1997.

Kuboniwa, Masaaki, and Shinichiro Tabata, eds., *Russian Economy at the Crossroads: Transition to Market and SNA*. Tokyo, Japan: Aoki Shoten, 1999 (in Japanese).

Kuboniwa, Masaaki, Shinichiro Tabata and Nataliya Ustinova, "How Large Is the Oil and Gas Sector of Russia? A Research Report," *Eurasian Geography and Economics*, **46**, 1:68-76, 2005.

Nakamura, Yasushi, "Extractive Industries and 'Dutch Disease,'" *Discussion Paper in Economics*, 99/5, Department of Economics, Heriot-Watt University, 1999.

Popov, Vladimir, "Uroki valiutnogo krizisa v Rossii i v drugikh stranakh," *Voprosy ekonomiki*, 6:100-122, 1999.

Shiobara, Toshihiko, *Economic Structure of Present Russia*. Tokyo, Japan: Keio University Press, 2004 (in Japanese).

Sistema tablits "Zatraty-Vypusk" Rossii. Moscow, Russia: Rosstat, various years.

Sotsial'no-ekonomicheskoe polozhenie Rossii (SEP). Moscow, Russia: Rosstat, monthly.

Tabata, Shinichiro, "Changes in the Structure and Distribution of Russian GDP in the 1990s," *Post-Soviet Geography and Economics*, **37**, 3:129-144, 1996.

Tabata, Shinichiro, "The Investment Crisis in Russia: A Research Report," *Post-Soviet Geography and Economics*, **38**, 9:558-566, 1997.

Tabata, Shinichiro, "Transfers from Federal to Regional Budgets in Russia: A Statistical Analysis," *Post-Soviet Geography and Economics*, **39**, 8:447-460, 1998.

Tabata, Shinichiro, "The Great Russian Depressions of the 1990s: Observations on Causes and Implications," *Post-Soviet Geography and Economics*, **41**, 6:389-398, 2000.

Tabata, Shinichiro, "Russian Revenues from Oil and Gas Exports: Flow and Taxation," *Eurasian Geography and Economics*, **43**, 8:610-627, 2002.

Tabata, Shinichiro, "Regional Sources of Federal Expenditure and the Pattern of Revenue Sharing in Post-Soviet Russia," *The Donald W. Treadgold Papers*, **36**:19-52, 2003.

Tabata, Shinichiro, "Russia's Economic Integration with CIS Countries," in Japan Research Institute, ed., *Report on the Survey of Russian Economic Reforms*. Tokyo, Japan: Japan Research Institute, 51-61, 2005.

Tabata, Shinichiro, "Price Differences, Taxes and the Stabilization Fund," in M. Ellman, ed., *Russia's Oil and Natural Gas: Bonanza or Curse?* London, UK: Anthem, 2006a.

Tabata, Shinichiro, "Observations on the Influence of High Oil Prices on Russia's GDP Growth," *Eurasian Geography and Economics*, **47**, 1:95-111, 2006b.

Tabata, Shinichiro, and Akihiro Iwashita, eds., *Slavic Eurasia's Integration into the World Economy and Community*. Slavic Eurasian

Studies, No. 2, Sapporo, Japan: Slavic Research Center, Hokkaido University, 2004 [http://src-h.slav.hokudai.ac.jp/coe21/publish/no2_ses/contents.html].

Tabata, Shinichiro, and Megumi Suezawa, eds., *The CIS: Restructuring of Former Soviet Space.* Tokyo, Japan: Kokusai Shoin, 2004 (in Japanese).

Uegaki, Akira, "Russian International Financing," in Vladimir Tikhomirov, ed., *Anatomy of the 1998 Russian Crisis.* Melbourne, Australia: Contemporary Europe Research Centre, University of Melbourne, 1999.

Uegaki, Akira, "Moscow and the Central Economic Area: Analysis of the Lack of Linkage," in Klaus Segbers, ed., *The Political Economy of Regions, Regimes and Republics: Explaining Post-Soviet Patchworks,* Vol. 3. Aldershot, Hants, England; Burlington, VT: Ashgate, 2001.

Uegaki, Akira, "Inostrannye investitsii v rossiiskikh regionakh," *Ekonomicheskaia nauka sovremennoi Rossii*, **2**:115-132, 2002.

Uegaki, Akira, "An Analysis of Russia's Embryonic Globalization: Regional Foreign Trade and Hard Currency Receipts," *The Donald W. Treadgold Papers*, **36**:55-75, 2003.

Uegaki, Akira, "Russia as a Newcomer to the International Financial Market, 1992–2000," *Acta Slavica Iaponica*, **21**:23-46, 2004a [http://src-h.slav.hokudai.ac.jp/index-e.html].

Uegaki, Akira, "Russia and the IMF," in Tabata and Iwashita (2004), 61-89, 2004b.

Uegaki, Akira, *Russia under Economic Globalization.* Tokyo, Japan: Nihon Hyoronsha, 2005 (in Japanese).

Table 1. Value Added Produced by Oil and Gas (in percent of GDP)

	1995	1996	1997	1998	1999	2000	2001	2002
Oil and gas total	15.82	17.52	16.07	14.80	19.60	24.1	20.5	18.9
Oil extraction products						11.6	10.0	9.3
Oil processing products						5.4	4.6	4.5
Gas sector products						7.1	5.8	5.1
Producers	5.16	5.46	5.38	4.99	6.49	7.8	6.7	6.5
Oil extraction products						5.7	5.1	4.8
Oil processing products						1.1	0.9	0.8
Gas sector products						1.0	0.7	0.9
Transport enterprises	3.79	3.87	2.77	2.33	0.98	1.0	1.1	0.9
Oil extraction products						0.5	0.5	0.4
Oil processing products						0.4	0.4	0.4
Gas sector products						0.1	0.2	0.1
Trade and intermediation enterprises	4.79	4.56	4.76	4.66	8.62	10.7	7.7	7.4
Oil extraction products						3.9	2.7	3.0
Oil processing products						2.5	1.9	1.9
Gas sector products						4.3	3.1	2.5
Net taxes on production	2.07	3.64	3.16	2.82	3.52	4.6	5.0	4.0
Oil extraction products						1.5	1.7	1.1
Oil processing products						1.4	1.4	1.3
Gas sector products						1.7	1.9	1.6

Sources : Data obtained by a joint project by Masaaki Kuboniwa and the Rosstat.

Table 2. Trade Margins of Oil and Gas (in billion rubles)

		1995	1996	1997	1998	1999	2000	2001	2002
Total	Oil and gas	101.6	128.3	154.9	161.7	552.3	997.3	942.7	1,154.7
	Oil extraction						342.4	315.7	451.8
	Oil processing						247.5	239.8	304.1
	Gas						407.4	387.2	398.8
Export margins	Oil and gas	54.9	86.6	91.8	98.9	417.4	650.4	473.1	575.4
	Oil extraction						221.6	174.8	327.0
	Oil processing						78.4	17.5	18.4
	Gas						350.4	280.8	230.0
Domestic margins	Oil and gas	46.7	41.7	63.1	62.8	134.9	346.9	469.6	579.3
	Oil extraction						120.8	140.9	124.8
	Oil processing						169.1	222.3	285.7
	Gas						57.0	106.4	168.8
Share of export margins (in percent)	Oil and gas	54.0	67.5	59.3	61.2	75.6	65.2	50.2	49.8
	Oil extraction						64.7	55.4	72.4
	Oil processing						31.7	7.3	6.1
	Gas						86.0	72.5	57.7

Sources: Compiled by the author from *Sistema*, various years.

Fig. 1. Value Added Produced by Oil and Gas and World Oil Prices

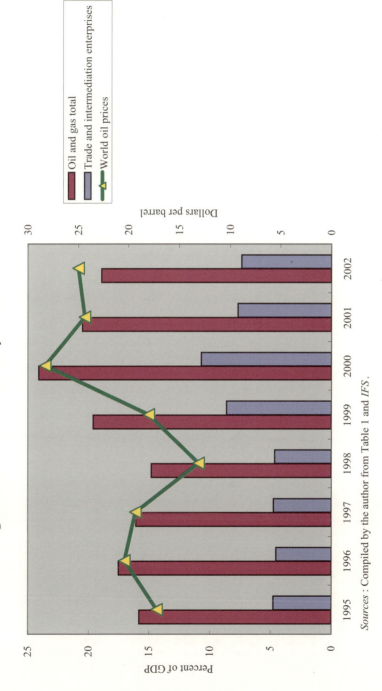

Legend:
- Oil and gas total
- Trade and intermediation enterprises
- World oil prices

Dollars per barrel (right axis): 0, 5, 10, 15, 20, 25, 30

Percent of GDP (left axis): 0, 5, 10, 15, 20, 25

Years: 1995, 1996, 1997, 1998, 1999, 2000, 2001, 2002

Sources : Compiled by the author from Table 1 and *IFS*.

Fig. 2. Real Exchange Rate of the Ruble (1992 Q2=100)

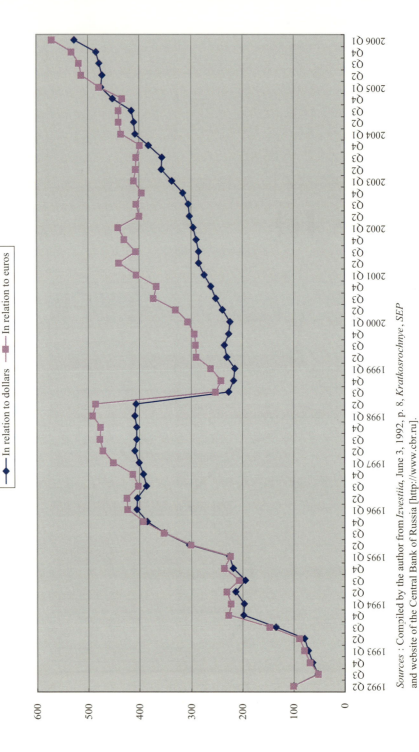

→ In relation to dollars → In relation to euros

Sources : Compiled by the author from *Izvestiia*, June 3, 1992, p. 8, *Kratkosrochnye, SEP* and website of the Central Bank of Russia [http://www.cbr.ru].

How Large is the Oil and
Gas Sector of Russia?
A Research Report[*]

Masaaki Kuboniwa
Shinichiro Tabata
Nataliya Ustinova

INTRODUCTION

It is difficult to overstate the importance of the Russian oil and gas sector, both to the Russian economy and to world hydrocarbon markets. Possessing 6 percent of proved world reserves of oil and 27 percent of natural gas, Russia accounted for 9 percent of global oil exports and 29 percent of gas exports in 2003.[1] Major customers included the European Union, China, and Japan, with considerable post-9/11 attention to the prospects for penetrating the U.S. market (e.g., see Aron, 2002; Butler, 2002). And internally, in the words of Clifford Gaddy (2004, p. 346), "It is becoming increasingly clear that Russia's oil sector has been and will for the foreseeable future continue to be the key to the country's economic performance." That being said, there is a considerable range of views as to the size of the sector, and there is a need to assess more critically what is actually being measured in the official statistics.

The World Bank (2004a, 2004b) only recently concluded that the share of oil and gas sector in Russian GDP was underestimated in the of-

---

[*] This paper was reprinted from *Eurasian Geography and Economics*, **46**, 1:68-76, 2005.

[1] All figures are derived from the British Petroleum 2004 Statistical Review of World Energy (BP, 2005) and Russian Oil (2005); Russia's corresponding world shares of oil and natural gas production were 11 and 22 percent, respectively.

ficial GDP statistics compiled by Goskomstat Rossii (Federal State Statistics Service of Russia) due to the prevalence of the transfer pricing.[2] In this brief paper we revisit the specific treatment of value added created in the oil and gas sector in Russian statistics and offer an alternative method of calculation based on the use of input-output tables, with "enterprise groups" serving as the units of statistical observation.

The official figure for the share of the oil and gas sector in Russian GDP can be derived only from the input-output tables compiled by Goskomstat Rossii. The most recent input-output tables available at present cover the year 2001 (*Sistema*, 2004). The problem with the official Russian figure is that it is very low, namely 7.8 percent in 2000 and 6.7 percent in 2001 (see Table 1). As discussed below, when we add a part of the value added attributed to the trade and transportation sectors (as trade and transportation margins and net taxes on oil and gas) to the official figure, we obtain substantially different figures: 24.1 percent in 2000 and 20.5 in 2001. If this is the case, the share of industry should be increased by some 10 percent, and the share of the trade sector should be reduced accordingly (here, we neglect net taxes on products). This outcome completely changes the structure of Russian GDP, and the contribution of the oil and gas sector to Russian economic growth must be reconsidered. We begin by outlining the relevant methodology employed by Goskomstat Rossii, and follow by presenting our alternative calculations and a comparison of the two methods.

THE METHODOLOGY OF GOSKOMSTAT ROSSII

Goskomstat Rossii's official SNA data are essentially based on international standards incorporated in the System of National Accounts (1993) as well as on Russian data detailing enterprise characteristics, prices, and employment. Value added of the oil and gas sector is recorded in the industries in which it is created, and is not "transferred" to the trading sector in order to inflate the weight of services in GDP.

---

[2] We should like to note here that this observation was made in the writings of Kuboniwa (2002, 2004a, 2004b) and Tabata (2002), as acknowledged by Sagers (2002), Ellman (2004), and OECD (2004, p. 20); preliminary observations were first published in Kuboniwa and Tabata (1999).

The issue here is not "transfer pricing" (a specifically Russian practice), but rather the presence of large holdings in the oil and gas sector, which include the following two types of enterprises: (1) producing enterprises that extract and process oil and gas; and (2) trading enterprises that sell the oil and gas in domestic and international markets. Both types are independent legal entities that generate their own statistical reports. Because the main activities of the first type comprise either extraction or processing, the value that they add is not large. The value by the second type (sales) is considerably larger than that of producing enterprises, because the gross profit of trading enterprises is the difference between international and domestic price levels. Thus, for example, in 2002 the average export price of gas (2,192 rubles per 1,000 cubic meters) was more than 11 times higher than the gas producers' price (194 rubles per 1,000 cubic meters). Such considerable price differentials accounted for the main income of the country's largest wholesale trading enterprise, Gazprom,[3] which for all practical purposes has been the exclusive exporter of natural gas since 1994.

Russia's enterprise holding groups include independent enterprise units that are legal entities engaged in a variety of activities. Each unit creates value added that is attributed to its main activity stipulated in the enterprise unit's registry. However, if one proceeds to classify all enterprises in a vertically integrated holding by the broad type of their industrial activity (e.g., as oil or gas), then the trading and intermediation activities will be materially underestimated. Moreover, the share of trade in the country's GDP will also decrease. In fact, such reclassification of trade and intermediation enterprises constitutes a redistribution of parts of value added from the trading sector to the oil and gas industries, in essence raising producers' prices for oil and gas. Thus, for example, if to the cost of natural gas produced in 2001 (63.3 billion rubles) one were to add the output of trading and intermediation enterprises that sell gas in the domestic and international markets (386.5 billion rubles), then the producers' price of gas would rise from 118 rubles per 1000 cubic meters (the official enterprise price statistic) to 839 per 1000 cubic meters. Hence, a contradiction will arise between price statistics tracking real producers'

---

[3] Gazprom has been registered as an economic unit in the foreign trade sector (Tabata, 2002, p. 612).

prices (and their dynamics), on the one hand, and the data of national accounts, on the other.

Moreover, value added is not an abstract indicator that can be moved from one industry to another. When calculated in terms of income generation, it represents real value formed by labor income and net taxes (taxes minus subsidies) on production, and gross profit. Thus, a transfer of the value added by trade to the oil and gas industries instantaneously raises the average wage of an oil or gas worker in Siberia several times. As a result, average wage and profitability indicators for enterprises in this industry will reflect highly distorted values, differing significantly from official labor statistics. And the industrial structure of the Russian economy would thus take on a "virtual" character that is unfamiliar to those who live and work in Russia.

There is also a regional aspect that deserves to be mentioned in this brief report. In addition to the country's GDP, Goskomstat Rossii calculates Gross Regional Product (GRP) values for each constituent entity of the Russian Federation. Inasmuch as most Russian holding groups (particularly the largest ones) are registered in Moscow and other large cities, most taxes are correspondingly collected and paid there. Thus a redistribution of value added among Russia's regions would be required to reflect the actual contribution of each region to the country's GDP.

Let us now have a look at specific calculations involving data based on input-output tables for the year 2001—the most recent statistics released by Goskomstat Rossii (*Sistema*, 2004). If one were to include in the oil and gas sector only that sector's production enterprises, the share of its value added to GDP, at market prices, would amount to 6.7 percent, as shown in Table 2. However, if one then includes the value added by enterprises that transport and sell oil and gas, as well as net taxes levied on the sector's output and paid into the state budget, the share rises to as much as 20.5 percent. Such calculations are routinely made at Goskomstat Rossii, which submits the resulting statistical data to ministries, agencies, and other interested users.[4]

---

[4] We similarly used data based on input-output tables in our calculations in the past (Kuboniwa, 2002, 2004a, 2004b; Tabata, 2002).

ALTERNATIVE CALCULATIONS

We now offer here an alternative method of calculating the contribution of the oil and gas sector to Russia's GDP. The method is based on a modification of the input-output tables (i.e., supply and use tables),[5] involving a different approach to the units of statistical observation. In the input-output tables for 2001, as well as in calculations of each industry's production, the unit of statistical observation is the "enterprise." An "enterprise" in this context is an organizational and legal entity producing output and enjoying a certain degree of autonomy in decision-making related to the distribution of resources at its disposal. However, because most if not all enterprises in the oil and gas sector are vertically integrated,[6] the "enterprise group" rather than the "enterprise" can be used as the statistical observation unit. Such an expanded unit incorporates a small number of organizational and legal entities, each constituting an independent legal nucleus involved in a different type of activity, all grouped together within a single legal or financial framework. Formally, although the enterprises in an enterprise group possess a certain degree of decision-making authority, they are in practice controlled by the headquarters of the enterprise group. Linked within a single vertically integrated technological chain, the enterprises become dependent on each other and lose the capacity to function independently.[7]

The aforementioned method allows us to modify the matrix of outputs of the supply table so that sales and specialized transportation, which support the shipping and marketing of the sector's products, are treated as secondary activities in oil extraction, oil processing, and gas industries. It follows that corresponding modifications also need to be made in the use table; Table 3 presents a fragment of the modified supply table for the year 2001.

We can now attribute the following to the oil-extracting sector: (1) a part of the output of wholesale, external trade, and intermediation activity

---

[5] The supply and use tables of the input-output system are explained in *System* (1993, pp. 351-361).

[6] As for the vertically integrated companies in the oil industry, see Dienes (2004, pp. 322-324).

[7] For example, oil and gas cannot be sold "in the ground" and must first be extracted and transported.

(corresponding to the volume of trade and intermediation price margin on oil); and (2) the output of oil pipeline transportation.[8] We can also attribute the following to the oil-processing sector: (1) a part of the output of wholesale, external, and retail trade, as well as a part of intermediation activity (corresponding to the amount of the trade and intermediation margin on oil products);[9] and (2) the output of pipeline transportation of oil products.[10]

Finally, within the framework of our approach, one can attribute to the gas sector: (1) a part of the output of wholesale and external trade, and of the intermediation activity (corresponding to the amount of the trade and intermediation margin on gas); (2) the output of gas transportation enterprises; and (3) the revenue from renting the pipeline (owned by Gazprom) to transportation enterprises.[11]

The analysis of the structure of the sector's output presented in Table 3 indicates that the share of trading and intermediation services (which are

---

[8] Oil transportation is controlled by AK Transneft', a large monopoly with 12 regional subsidiaries. As an essential component in the delivery to domestic and international users, Transneft' handles pipelines through which the flow of exports generates the lion's share of revenues of the major oil enterprises. The pipeline transportation margin amounts to ca. 80 percent of the total transportation margin on oil, with the balance divided among railway, marine, and, marginally, highway transport. However, the output of these transportation sectors (as measured by the transportation margin on oil) is not included in the output of the vertically integrated structure of the oil-extracting sector. For example, railway transportation is handled by another large Russian monopoly, Russian Railways (RZhD), whereas marine transport is in the hands of independent shipping enterprises. Moreover, these transportation industries handle a variety of cargo in addition to products of the oil sector.

[9] Attribution of a part of retail trade (corresponding to the amount of the retail margin on oil products) to the sector is due to the ownership of fuel-retailing enterprises (gas stations, etc.) by giants such as Lukoil, Sibneft', Slavneft', and still in part by the now largely dismantled Yukos.

[10] As in the case of oil, railroad, marine, domestic shipping, and highway transportation (as measured by the margin on oil products) are not attributed to the oil processing sector.

[11] It should be noted that a part of the retail trade conducted outside of the Gazprom system has not been attributed to the gas sector (i.e., the retail margin on gas processing products, namely 708.2 mln. rubles). It should also be called to the reader's attention that although a part of railway transportation (the gas price margin) was not attributed to the gas sector, pipeline transportation services (220.1 billion rubles) exceed the entire transportation margin on gas (28.4 billion rubles), because the bulk of these services is neither treated nor recorded as a margin, but rather as intermediation input costs incurred by wholesale and external trade enterprises engaged in the selling of gas.

essentially secondary types of activity) in the oil-extracting and processing industries amounts to around 30 percent of their output. At the same time, the share of such services in the gas sector (at 47.6 percent) is almost four times higher than the output of the sector's main activity (i.e., extraction and processing at 12.1 percent). From the perspective of SNA theory, such a modified output matrix may look unusual. One should remember, however, that it does reflect the realities of the Russian economy.

After appropriate modifications of the supply table, we also made changes in the use table.[12] The structure of input consumed by the oil and gas industries was calculated by using data on individual enterprises as well as data on corresponding transport and trade industries derived from input-output tables. Value added was determined as the difference between the output and intermediate input consumed by these industries. Table 4 shows the value added by industries of the oil and gas sector, calculated on the basis of modified input-output tables for the year 2001. The table indicates that most of the value added to oil extraction sector is created by industrial producers, while in the oil processing and gas sectors, the bulk is added by trade and intermediation activities.

COMPARING THE TWO METHODS

We can now compare the shares of value added in Russia's GDP that are based on official input-output tables with those based on modified tables. The results are presented in Table 5. As one can readily ascertain, the GDP's share of value added to the oil and gas sector in basic prices is 10.9 percent higher in the modified input-output tables than in the official ones. The increase is largely due to the inclusion in the modified tables of trade and intermediation activities (8 percent of the increase), and to a relatively minor extent, to the inclusion of transportation (1.7 percent) and real estate operations (1.1 percent).

In Table 6 we present the change (in basic prices) in the structure of GDP across all industries of the Russian economy caused by modification of input-output tables.[13] The table shows that even after our modification

---

[12] More specifically, parts of the output and intermediation inputs in the trading and intermediation sectors as well as those in transportation and real estate operations were added to the corresponding industries comprising the sector.

[13] While the modification based on the notion of enterprise groups is presented here

of the input-output tables, the share of trade and intermediation activities in GDP (in basic prices) remains high, amounting to 21.2 percent.

It should be noted that there is no difference between both sets of tables in 2001 in the value of trade and transportation margins and in other components of purchasers' prices for products of the oil and gas sector. Thus, Table 7, detailing purchasers' prices for oil and its products, shows that they are almost twice as high as their basic prices. For natural gas, the purchasers' prices are more than six times as high as the basic. All such considerable differences between the purchasers' and basic prices are due to the size of the trade and intermediation margins and taxes[14] on these types of products.

CONCLUDING NOTE

The alternative calculation presented in this research report sheds some additional light on specific features of the oil and gas sector's contribution to Russia's GDP. The sector's uniqueness is rooted not only in its magnitude and undisputed importance, but also in the presence and role of enterprise groups that command and control its workings. We argue that these specific characteristics, rather than faulty methodological treatment by Goskomstat Rossii, are largely responsible for the material discrepancy between official statistics and the figures discussed above.

REFERENCES

Aron, Leon, "Russian Oil and U.S. Security," *The New York Times*, May 5, 2002, WK15.

BP (British Petroleum), *BP 2004 Statistical Review of World Energy* [http://www.bp.com/subsection.do?/categoryID=95&contentID=2006 480], accessed January 23, 2005.

Butler, Richard, "A New Oil Game, with New Winners," *The New York Times*, January 18, 2002, A25.

---

solely for the oil and gas sector, similar problems exist (albeit considerably less pronounced) in such sectors as aluminum and steel.

[14] Export, excise, and value-added taxes.

Dienes, Leslie, "Observations on the Problematic Potential of Russian Oil and the Complexities of Siberia," *Eurasian Geography and Economics*, **45**, 5:319-345, 2004.

Ellman, Michael, "Russian Economic Boom, Post-1998," *Economic and Political Weekly*, July 17, 2004, 3234-3237.

Gaddy, Clifford, "Perspectives on the Potential of Russian Oil," *Eurasian Geography* and *Economics*, **45**, 5:346-351, 2004.

Kratkosrochnye ekonomicheskie pokazateli Rossiiskoi Federatsii. Moscow, Russia: Rosstat, monthly.

Kuboniwa, Masaaki, "An Analysis of Singularities of Russia's Marketization Using Input-Output Tables," *The Journal of Econometric Study of Northeast Asia*, **4**, 1:1-13, 2002.

Kuboniwa, Masaaki, "An Analysis of Expansion of the Trade Service Sector in Russia," in E.G. Iasin, ed., *Modernizatsiia ekonomiki Rossii: sotsial'nyi kontekst: v chetyrekh knigakh*, Vol. 1. Moscow, Russia: Vysshaia shkola ekonomiki, 2004a, 322-339.

Kuboniwa, Masaaki, "Russia's Oil and Gas Reconsidered," in Shinichiro Tabata and Akihiro Iwashita, eds., *Slavic Eurasia's Integration into the World Economy and Community*. Slavic Eurasian Studies, No. 2, Sapporo, Japan: Slavic Research Center, Hokkaido University, 2004b, 143-159 [http://src-h.slav.hokudai.ac.jp/coe21/publish/no2_ses/contents.html].

Kuboniwa, Masaaki, and Shinichiro Tabata, eds., *Russian Economy at the Crossroads: Transition to Market and SNA.* Tokyo, Japan: Aoki Shoten, 1999 (in Japanese).

OECD (Organisation for Economic Cooperation and Development), *Economic Surveys, Russian Federation.* Paris, France: OECD, 2004 [http://www.oecd.org/home/].

Osnovnye pokazateli sistemy natsional'nykh schetov. Moscow: Rosstat, 2004 [http://www.gks.ru/].

"Russian Oil Exports Up in 2004," *The Russia Journal*, January 22, 2005 [http://www.russiajournal.com/news/cnews-article.shtml?nd=46971], accessed January 23, 2005.

Sagers, Matthew J., "Comments on the Flow and Taxation of Oil-Gas Export Revenues in Russia," *Eurasian Geography and Economics*, **43**, 8:628-631, 2002.

Sistema tablits "Zatraty-Vypusk" Rossii za 2000 god. Moscow, Russia: Goskomstat Rossii, 2003.

Sistema tablits "Zatraty-Vypusk" Rossii za 2001 god. Moscow, Russia: Rosstat, 2004.

System of National Accounts, 1993. Brussels/Luxembourg/New York/ Paris/Washington, DC: CEC, IMF, OECD, United Nations, World Bank, 1993.

Tabata, Shinichiro, "Russian Revenues from Oil and Gas Exports: Flow and Taxation," *Eurasian Geography and Economics*, **43**, 8:610-627, 2002.

World Bank, *Russian Economic Report,* no. 7, 2004a [www.worldbank .org.ru].

World Bank, *From Transition to Development: A Country Economic Memorandum for the Russian Federation,* Draft, 2004b [www .worldbank.org.ru].

Table 1. Value Added at Basic Prices (percent of total GDP at market prices)

Component	2000	2001	2002	2003	2004
Industry	28.0	25.3	24.5	23.9	24.9
Oil and gas sector	7.8	6.7			
Transportation and communications	8.0	8.0	8.1	8.0	7.7
Transportation margins of oil and gas	1.0	1.1			
Trade and catering	21.2	19.9	19.9	19.9	19.5
Trade margins of oil and gas	10.7	7.7			
Net taxes on products	11.4	12.3	11.5	11.9	
On oil and gas	4.6	5.0			
Total contribution of oil and gas sector	24.1	20.5			

Sources: Compiled by the authors from *Osnovnye*, 2004, *Kratkosrochnye*, No. 11, 2004, *Sistema*, 2003, 2004, and unpublished Goskomstat Rossii data.

Table 2. Share of Oil and Gas Sector Output in Aggregate GDP, 2001 (in percent of GDP)

Components	Oil extraction products	Oil processing products	Gas sector products	Total
Total	10.0	4.6	5.8	20.5
Producers	5.1	0.9	0.7	6.7
Transport enterprises	0.5	0.4	0.2	1.1
Trading and intermediation enterprises	2.7	1.9	3.1	7.7
Net taxes on products	1.7	1.4	1.9	5.0

Sources: Compiled by the authors from *Sistema*, 2004 and unpublished Goskomstat Rossii data.

Table 3. Fragment of the Modified Supply Table for 2001

Product and services	Oil extraction sector		Oil processing sector		Gas sector	
	Mill. rubles	Pct. of total	Mill. rubles	Pct. of total	Mill. rubles	Pct. of total
Oil extraction products	647,458.2	62.3	—	—	591.7	0.1
Oil processing products	2,159.0	0.2	572,889.7	69.4	12,989.5	1.6
Gas industry products	3,312.2	0.3	167.3	0.0	98,366.7	12.1
Other industrial products	3,749.3	0.4	11,007.9	1.3	3,213.2	0.4
Pipeline transport services	67,520.3	6.5	1,706.6	0.2	220,092.7	27.1
Trading and intermediary services	315,692.8	30.4	239,815.0	29.0	386,478.6	47.6
Real estate services	—	—	—	—	89,841.5	11.1
Total	1,039,891.8	100.0	825,586.5	100.0	811,573.9	100.0

Sources: Compiled by the authors from unpublished Goskomstat Rossii data.

Table 4. Value Added of the Oil and Gas Sector Industries in 2001

Components	Oil extraction sector		Oil processing sector		Gas sector	
	Mill. rubles	Pct. of total	Mill. rubles	Pct. of total	Mill. rubles	Pct. of total
Industrial producers	453,963.1	60.6	78,815.1	29.6	64,070.4	14.6
Pipeline transportation	31,033.9	4.1	810.0	0.3	105,160.0	24.0
Trade and intermediation activities	264,534.8	35.3	186,253.3	70.1	179,729.8	41.0
Real estate operations	—	—	—	—	89,841.5	20.5
Total	749,531.9	100.0	265,878.4	100.0	438,801.7	100.0

Sources: Compiled by the authors from unpublished Goskomstat Rossii data.

Table 5. Value Added in Basic Prices Created by Industries of the Oil and Gas Sector in 2001 (in percent of GDP)

Industry/sector	Producers	Transporta-tion	Trade and intermedi-ary services	Real estate	Total
	Official input-output tables				
Oil extraction	5.8	—	—	—	5.8
Oil processing	1.0	—	—	—	1.0
Gas	0.8	—	—	—	0.8
Oil and gas sector	7.6	—	—	—	7.6
	Modified input-output tables				
Oil extraction	5.8	0.4	3.4	—	9.5
Oil processing	1.0	0.0	2.4	—	3.4
Gas	0.8	1.3	2.3	1.1	5.6
Oil and gas sector	7.6	1.7	8.0	1.1	18.5

Sources: Compiled by the authors from *Sistema*, 2004 and unpublished Goskomstat Rossii data.

Table 6. Structure of GDP in Basic Prices for 2001 (in percent of total)

Sector	Official input-output tables (a)	Modified input-output tables (b)	Difference (b–a)
Electric power	2.8	2.8	—
Oil extraction	5.8	9.5	3.8
Oil processing	1.0	3.4	2.4
Gas industry	0.8	5.6	4.8
Coal industry	0.5	0.5	—
Other fuel industries	0.0	0.0	—
Ferrous metallurgy	1.5	1.5	—
Nonferrous metallurgy	3.1	3.1	—
Chemical and petrochemical industry	1.5	1.5	—
Machine-building and metalworking	4.7	4.7	—
Forestry and wood-processing[a]	1.2	1.2	—
Construction materials industry[b]	0.9	0.9	—
Light industry	0.5	0.5	—
Food industry	3.6	3.6	—
Other industries	0.7	0.7	—
Industry, total	28.7	39.6	10.9
Construction	7.5	7.5	—
Agriculture and forestry	6.9	6.9	—
Transportation and communications	9.6	7.9	-1.7
Trade, intermediation, and catering	30.3	21.2	-9.1
Other goods and services production	0.8	0.8	—
Housing maintenance and other services[c]	2.9	2.9	—
Medicine, education and culture[d]	5.6	5.6	—
Science and scientific services[e]	1.5	1.5	—
Finance, credit, insurance, and administration[f]	7.9	7.9	—
Financial intermediation services[g]	-1.6	-1.6	—
Total in basic prices	100.0	100.0	0.0

[a]Including pulp and paper.
[b]Including glass and porcelain production.
[c]Including non-productive services for individual consumers.
[d]Including social services and sports.
[e]Including geology, resource prospecting, land surveying, and hydrometeorology.
[f]Including social organizations.
[g]Indirectly measured.
Sources: Compiled by the authors from *Sistema*, 2004 and unpublished Goskomstat Rossii data.

Table 7. Purchasers' Prices for Products of the Oil and Gas Sector in 2001 (in percent)

Commodity	Basic price	Trade and intermediation margin	Transport margin	Net taxes on products	Purchasers' price
Oil	100.0	47.7	13.0	23.3	184.0
Oil processing products	100.0	37.6	8.4	19.8	165.8
Gas	100.0	372.1	27.3	164.1	663.6

Sources: Compiled by the authors from *Sistema*, 2004, p. 62.

Chapter 3

Economy-wide Influences of the Russian Oil Boom: A National Accounting Matrix Approach

Yasushi Nakamura

INTRODUCTION

The ongoing high price in the world oil market gives Russia large windfall export revenue. There is not much room to argue against the assertion that the high oil price has been contributing to the recent good performance of the Russian economy. At the same time, the menace of the "Dutch disease" has been discussed more and more intensively: the large trade surplus leads to appreciation of the ruble and brings additional funds to be spent. The ruble appreciation hampers export by the non-oil industries and makes imports cheaper. The additional funds may be spent on non-tradables and imports, while the demand for domestic tradables decreases. On the supply side, the stimulated oil and non-tradable sectors pull resources at the cost of the other export manufacturing sectors. In short, the Russian economy grows and turns into an "oil-monoculture" economy instead of a developed industrial and service economy. This "Dutch disease" scenario is undoubtedly one of the theoretically possible growth paths of Russia; we need, nevertheless, to carefully and comprehensively examine the economic relations between the oil sector and the Russian national economy before diagnosing the reality of the symptoms leading to the disease.

To explore all aspects of financial and real influences of an oil boom on a national economy is a complicated task. For Russia, the lack of economic data, the shortness of the time series, and the ongoing structural reforms increase the difficulty. In this chapter, we use a Russian national accounting matrix (NAM) instead of a full-fledged economy-wide model and limit our object to answering only one question: how much demand for domestic tradables can the Russian oil sector create? Our working hypothesis is that the Russian oil industry creates demands for domestic tradables as much as do the other sectors and, in this respect, the oil industry is able to contribute to growth of the Russian manufacturing sector. The other fundamental questions, that is, how much will the ruble be appreciated by the boom and what kind of real and financial effects will the appreciation bring, should be analyzed later, after constructing a Russian macroeconomic model including the financial and monetary sphere.

The input-output (I-O) methods have long been applied to analyze resource booms (see Davis, 1995, p. 1767). The conventional I-O methods are, however, not sufficient because these methods can neither consider the multiplier effects diffused through consumption and investment expenditures nor examine the economic influences of additional financial funds brought by the increasing oil export. Most importantly, they cannot take the price effects into account. These limitations are particularly serious for the present Russian oil boom, where a large part of the initial impact of the boom on the economy takes the form of an increase in financial funds denominated in foreign currency. To overcome the defect of the I-O analysis, we compile a Russian NAM and calculate the "accounting multipliers."

A NAM differs from a conventional input-output table (IOT) in that a NAM includes income and expenditure accounts of the institutional sectors; a NAM, therefore, can show production, income, and expenditure as a complete circular flow. By setting any number of the accounts in a NAM to be exogenous, we can calculate the accounting multiplier matrix, which is equivalent to the Leontief inverse matrix of the I-O analysis. Because consumption and investment usually have large weights in the economic flow of a national economy, endogenising income and expenditure flows is a clear advantage of the NAM method. Using the Russian NAM, we can estimate the influences of the oil boom on the national economy through the consumption and investment linkages. On the other hand, the above procedure implies that we assume linearity of consumption and investment behaviors and ignore the effects of changes in the prices and the exchange rate. The linearity assumption hardly holds true, and the ignorance of price changes reduces the value of the NAM method. This method is, nevertheless, useful for studying the structural features of an economy and may be the only feasible method under the present circumstances to investigate economy-wide influences of the Russian oil boom.

The rest of the chapter is organized as follows: Section 2 gives a brief survey of the Dutch disease economics and the resource curse theory to confirm why we need to be cautious to directly apply the general theory to a particular resource boom. Section 3 explains the NAM methods and examines the data. Section 4 reports the results followed by a discussion of those results.

CONSIDERATION OF DUTCH DISEASE
AND RESOURCE CURSE

The role of extractive industries in economic development has long been discussed, receiving mixed evaluations. From a historical point of view, the extractive sector seems to have contributed to the Industrialization of many developed economies, including that of Japan. Davis (1995) reports that developing economies with significant extractive industries recorded better growth performance than that of other developing economies in the period of 1970 to 1990. Askari and Jaber (1999) admit that there were negative influences of the oil boom in the 1970s on economic development of the oil-exporting countries of the Persian Gulf; they, nevertheless, concluded that the countries made great strides in terms of enhancing the overall welfare of their citizens during the period. There seems to be little room to argue that Indonesia, which has a large extractive sector, has recorded an impressive growth of the export-oriented manufacturing sector during the last thirty years (see Usui, 1996; Rodgers, 1998).

On the other hand, the "Dutch disease" economics suggests that extractive industries may contribute to economic development not only at a lower level than expected, but also negatively. The Dutch disease economics tends to more or less relate negative influences of an expanding extractive industry to large and sudden economic shocks such as the natural gas boom in the Netherlands and the oil shocks in the 1970s. The "resource curse" thesis further suggests that it is normal and usual for the extractive sector to impede economic growth. Discussions of the resource curse appear in Amuzenger (1982), Gelb (1986, 1988), Auty (1994),

Davis (1995), and Sachs and Warner (1995, 2001). Their arguments are as follows.

First, the Dutch disease economics suggests that the expanding extractive sector can adversely affect the manufacturing sector mainly through real exchange rate appreciation (increase in relative price of non-tradable to tradable goods).

Second, assuming that "learning by doing" is one of the most important factors to increase manufacturing productivity, and that the effectiveness of learning by doing depends on the volume of the activity, even a temporal contraction of manufacturing production can result in an irrevocable loss of competitiveness. A resource boom which impedes the development of the manufacturing sector is, therefore, highly undesirable.

Third, extractive industries usually generate large rents, most of which go to the government. This affluence tends to induce rent-seeking activities, mismanagement of the public fund such as over-ambitious public investment projects, and lax social and economic policies.

Fourth, the extractive sector is supposed to induce little demand for domestic tradables. Most mineral income may be spent on non-tradables through government spending, or repatriated. The extractive sector can be regarded as an "enclave" in an economy in terms of demand creation of tradables (see Bosson and Bension, 1977), while it may induce excessive demand for non-tradables.

Finally, volatile and often violent changes in the conditions of mineral production and marketing might cause economic and political problems, particularly when a mining recession forces the government to tighten its lax policies. Moreover, the eventual depletion of mineral deposits may cause large structural adjustment costs.

These arguments seem more or less valid in the light of historical experiences of oil-exporting developing economies during and after the two oil shocks in the 1970s, analyses of which the resource curse thesis is mostly based on. It is, however, not clear to what extent the historical experiences can be accounted for by the general characteristics of the extractive sector and to what extend by other particular factors in each case. From this perspective, the following points should be noted.

First, the theoretical framework of the Dutch disease economics, which is often referred to as the "core model," is undoubtedly valid. The problem addressed by the core model is, however, not a property of the extractive sector, but adjustment problems accompanying structural changes in general. The core model explains Dutch disease as follows (see Corden, 1984; Wijnbergen, 1984; Neary and Wijnbergen, 1986b; Sitz, 1986). A booming extractive industry pulls production factors and resources (resource movement effect) and increases income mostly through export of the mineral products. The increased mineral income is thought to expand demand for non-tradables, mostly through government spending (spending effect). The increase in demand for non-tradables raises relative prices of non-tradables (real exchange rate appreciation effect). Consequently, non-tradable sectors pull more resources, and fewer resources are available for non-extractive tradable sectors. This causal relationship obviously holds true not only for a mineral boom but also for any expansion of an export-orientated tradable sector, although magnitudes of the effects may differ among the extractive and the other tradable sectors.

Second, volatility and unpredictability are inherent in extractive industries; they are, however, not exclusive to the extractive industries. If the future course of a resource boom could be perfectly foreseen, Dutch disease effects would be no more than rational changes in the economic structure. The changes would claim adjustment costs and would be brought in only when the costs could be covered with the returns from the changes. The future is, however, not perfectly foreseeable in the real world. The changes could cause adjustment costs that would not bring any benefits. It might be true that uncertinities of extractive activities are so intensive that only an extractive sector can cause irrational structural changes and infertile adjustments costs to such an extent that the oil shocks have done. The negative influences of the property of the extractive sector, nevertheless, seem manageable in the long run. If this were not the case, the extractive sector would have disappeared long ago.

Finally, it is certain that problems of efficient and effective distribution of mineral rents and spending of government revenue arise on a large scale if the boom increases them on a large scale. Moreover, a large amount of mineral rents may tempt people to undertake rent-seeking activities and even criminal activities. Bad governance and mismanagement of government spending are, however, neither a problem inherent in the extractive sector nor a problem caused by the extractive sector only. We are certainly able to find a number of cases of bad governance among countries that have never experienced any export boom.

In summary, it would not be very fruitful to evaluate a particular case of export boom based on the general theory of "Dutch disease" and "resource curse." Before concluding our study of the outcome of a resource boom, we need to carefully analyze the economic relations of the booming sector to the national economy in the conditions specific to the economy.

DATA AND METHOD

To analyze the economy-wide influences of the Russian oil boom, we use a NAM, which can be regarded as an extended IOT including endogenous income and expenditures flows (see Table 1). The 2001 Russian NAM was compiled, based on the 2001 IOT (*Sistema*, 2004) and the 2001 national accounts (*Natsional'nye*, 2004). Nakamura (2004) explains the details of the compilation procedure. Nakamura (2004) refers to a 1999 Russian NAM, but the frameworks of the 1999 and 2001 NAMs are almost identical, except that the 2001 NAM does not have the institutional accounts separated for the oil and gas companies. Unlike the case of 1999, discrepancies between the IOT and the national accounts were small in 2001. If we move the amount equal to the secret wage in the 2001 national accounts from the gross operating surplus in the 2001 IOT to the labor income in the 2001 IOT, most discrepancies between the IOT and the national accounts disappear. This may simply reflect the fact that the 2001 IOT was updated from the 1995 benchmark IOT with some 2001 data (*Sistema*, 2004, p. 4). Discrepancies and inconsistencies remained nevertheless, and they were eliminated by a mathematical adjustment method (see Nakamura, 2004, p. 159).

We also compiled NAMs of the four OECD-member countries, namely, Australia, Canada, the Netherlands, and the United Kingdom, in the period of 1970–1990 to use them as a yardstick against the Russian NAM. The NAMs for Russia and the four countries have a common framework (see Table 1); however, it is not possible to directly compare the Russian NAM with the NAMs of the four OECD countries because of the methodological differences. The NAMs for the four OECD countries were constructed from the IOTs at current prices, which were included in the OECD I-O database (OECD, 1995), and the national accounts (UN, various years). Nakamura (1999) explains the details of the compilation procedure and the data. The four countries were chosen from ten countries included in the OECD I-O database, because they had a relatively large extractive sector (see Table 2). The extractive sector in the NAMs of the four OECD countries corresponds to code 2 of ISIC rev. 2, "mining and quarrying." For the Russian NAM, it is an aggregated sector of oil drilling and gas mining. The oil refinery sector, which is included in the "fuel-energy sector" of the Russian industry classification, is a manufacturing sector in our analysis as it is by the ISIC. Both for Russia and for the four OECD countries, the tradable sectors correspond to manufacturing plus electricity, while the non-tradable sectors include services and utility industries such as construction, trade, communication, and transportation. The Russian NAM includes the Use table (products by sectors table) and the Make table (sectors by products table), while the NAMs of the four OECD countries include symmetric IOTs (products by products table). This is, however, not a serious obstacle, because we can calculate influences of a change in a production account on other production accounts easily in the Russian NAM. Hereafter, we use the terms of "sector" and "product (or goods)" interchangeably for simplicity.

The NAMs were analyzed as follows.

First, the direct input structure of the extractive sector was examined. The input coefficients of labor (L), capital (C), tradables (T), non-tradables (N), and imports (M) of the extractive (R) sector were compared with those of the average tradable (T) and non-tradable (N) sectors to identify technological characteristics of the extractive sector. The capital (C), or capital income, is defined as the sum of gross operating surplus; it is difficult to compare the capital shares between the countries because of different treatments of indirect taxes.

Second, the standard sensitivity and power of dispersion analysis was applied to the production accounts part of the Russian NAM to investigate the economy-wide influences conducted only through the production linkages.

Third, the accounting multipliers of the NAMs were analyzed. The accounting multiplier matrix corresponds to the Leontief inverse matrix in the I-O analysis and the accounting multipliers to the complete input coefficients. In the following analysis, the export and the financial transactions with the Rest of the World in the NAMs were exogenized to calculate the accounting multiplier matrix; Import was endogenized. The accounting multipliers in our NAM analysis can therefore be regarded as complete input coefficients of a "closed model." The accounting multipliers of the extractive sectors were compared with those of the tradable (T), non-tradable (N), and production (P) sectors. The accounting multipliers of those aggregated sectors were the simple average of the accounting multipliers of the sectors included in each aggregated sector.

RESULTS

Technological Features of the Extractive Sector

Table 3 compares the input structures of the extractive sector and the other production sectors. From Table 3, the following facts can be identified:

(1) **Labor and Capital.** The labor input coefficient of the extractive sector (E_{LR}) was smaller than that of the tradable (T) and the non-tradable (N) sectors (E_{LT} and E_{LN}, respectively) except for $U68$ and E_{LT} of $A68$. The Netherlands cases were excluded because the Netherlands NAMs did not show labor and capital separately. On the other hand, the capital input coefficient of the R sector (E_{CR}) was larger than that of the other sectors (E_{CT} and E_{CN}) except for E_{CN} of $U68$, $R01b$, and $R01p$; the differences were small in the Russian cases. Comparison of the capital input coefficient between the R sector and N sector did not show a clear tendency. Regarding the total value added (labor and capital income), the share of total value added of the R sector was larger than that of the T sector for all cases.

The input coefficients of labor and capital (E_{LR} and E_{CR}) of the extractive industry seem to be smaller in Russia than in the four OECD countries. There is no OECD case where the labor income share is smaller than that in Russia. For the capital income share, $U68$ is the only clear opposite case, and $C71$ and $C90$ are the vague cases.

(2) **Domestic Tradables and Non-tradables.** The extractive (R) sector used fewer domestic intermediates per unit of output than the tradable (T) sector except for the non-tradable (N) inputs in the three Canadian cases and the $R01b$ case. In comparison between the R and N sectors, the input coefficients of domestic intermediates of the R sector seem to be smaller than those of the N sector.

(3) **Imports.** The input coefficient of imports (M) for the R sector (E_{MR}) was small in comparison with that of the T sector (E_{MT}) for all cases. The small E_{MR} apparently reflects that the R sector uses relatively few tradables as intermediate inputs. On the other hand, E_{MR} is at a similar level of the input coefficient of imports for the N sector (E_{MN}).

(4) Private Consumption and Gross Fixed Capital Formation (GFCF). The composition of private consumption and the GFCF seems to be common in the four OECD countries. Compared to the four OECD countries, the share of the non-tradables for Russia looked very small. It is, however, difficult to judge because of the large transaction costs, which consist of transportation and trade services. If we move the transaction costs, which account for the greatest part of the difference in the shares of the tradables between at basic prices and at purchaser prices ($31 = 50 - 19$) to the non-tradables, then the Russian share of the non-tradables in private consumption does not seem to be very small. One notable difference is that the share of imports in private consumption in Russia is larger than that in all four OECD countries.

The compositions of the GFCF expenditure had showed some tendency within a country but varied between the four OECD countries. For Russia, the share of imports in the GFCF expenditure seemed to be large in comparison with the share of the domestic tradables; the magnitude of the imports share is, however, not particularly large in comparison with the four OECD countries.

In summary, there seems to be little difference in the input and expenditure structures in Russia and the four OECD countries. The R sector uses less labor and fewer intermediates and generates more capital income than does the T sector in terms of unit of output. Nakamura (1999) identified that the input structures of the R sectors in the four OECD countries were highly similar after calculating correlation coefficients between the input coefficient vectors of the R sectors in the four OECD countries. The input structure of the R sector in Russia also seems to be not very different from that in the four OECD countries. On the other hand, the R sector was more or less similar to the non-tradable (N) sector in the input structure. The R and N sectors differed in the composition of value added: labor income was small and capital income was large for the R sector.

Economy-wide Influences of the Extractive Sector

Figure 1 indicates the result of the sensitivity and power of dispersion analysis using the production part of the Russian NAM. The figure shows the oil drilling, oil refinery, and gas mining sectors separately. Both indicators of the oil refinery sector were relatively high. The indices of the gas mining sector were almost at the average level. They may be a little smaller than those of the manufacturing sectors; most of the less than average sectors are non-manufacturing sectors. The sensitivity of the oil-drilling sector is higher than the average level, while its power of dispersion is significantly smaller than the average level. The small input coefficient for tradable intermediates seems to lead to the low power of dispersion of the oil-drilling sector. In summary, the Russian extractive sector creates less demand than the manufacturing sectors do, mostly because the extractive sector uses domestic tradables relatively little. It is arguable whether the differences of 10–20 percent in the power of dispersion index between the oil-drilling sector and the manufacturing sectors are economically significant. From the inverted point of view, we can say that the oil-drilling sector creates demand to 80–90 percent of the level of the manufacturing sectors.

The sensitivity and power of dispersion analysis concerns the demands induced through the production linkages. Table 4 summarizes the result of the accounting multiplier analysis, which enable us to examine the ripple effects defused through not only production linkages, but also income and expenditure flows. Table 4 indicates the following features of the economy-wide influences of the extractive (R) sector.

(1) Influences on Tradable (T), Non-tradable (N), and Production (P) Sectors. The R, T, and N sectors were not very different in their influences on the T sector. For $A68$, $U68$, and $R01$, the R sector influences were even stronger than the average in terms of the percentage deviation. In terms of the absolute values of the accounting multipliers, however, the differences do not seem significant for all cases. This finding may be supported by Benjamin, Devarajan, and Weiner (1989) and Fardmanesh (1990, 1991), where it is empirically and theoretically suggested that the extractive sector can stimulate manufacturing sectors through creating demands for domestic manufactured goods.

Regarding influences on the N sector, we can find the tendency clearly: The R sector and N sector influence the N sector more strongly than the T sector does. The difference between the R and N sectors does not seem significant. From this, we can conclude that the R sector influences the average production sector more strongly than does the T sector, and almost as much as the N sector. Moreover, this tendency can be seen commonly both in Russia and in the four OECD countries.

(2) Influences on Labor Income (*L*) and Capital Income (*C*). The *R* sector seems to influence labor income more strongly than does the *T* sector; *U90* is the sole exception. The *N* sector may influence labor income more than the *R* sector; whether this is so is not very clear. For capital income, it is clear that the influence of the *R* sector is stronger than not only the *T* sector but also the *N* sector in terms of percentage deviation. It is, however, arguable whether the differences are economically significant in terms of absolute values of the accounting multipliers.

In summary, the *R* sector influences the average production (*P*) sector as strongly as the *N* sector does and slightly more strongly than the *T* sector does. The difference between the *R* and *T* sectors in the influence on the *P* sector is accounted for by the fact that the *R* sector influences the *N* sector slightly more strongly than the *T* sector does. One reason for the strong influence of the *R* and *N* sectors on the economy is that their shares of imports in their expenditure are small. Because imports and net lending are the sole leak from the NAM system, the smaller share of imports in expenditures implies larger multiplier effects. Another reason is that the shares of labor and capital income are large in the *R* and *N* sectors. Most labor and capital income is spent on consumption and investment, and the shares of non-tradable goods are relatively large for both household consumption and investment (see Table 3). Expanding *R* and *N* sectors, therefore, tend to induce more demands for non-tradable goods than an expanding *T* sector in the first stages of the multiplier process. Because the input coefficient of imports for the *N* sector is small and the input coefficient of non-tradable goods for the *N* sector is large, the *N* sector induces lasting and large multiplier effects when it is once stimulated.

Another notable point is that the extractive industries in Russia and the four OECD countries showed similar tendencies. Nakamura (1999) showed that the patterns of economy-wide influences in the four OECD countries are very similar between not only the *R* sectors, but also between the *R*, *T*, and *N* sectors, regardless of the years and the countries. The Russian NAM seems to share this feature. The reason for this phenomenon is the dominancy of household consumption and fixed investment in our NAM model. The share of labor income in production cost is large for any sector; it is also usual that a large part of capital income, another large component of production cost, flows to households through property income flows. Household consumption, therefore, played a decisive role in shaping the pattern of economy-wide influences, no matter which sector an exogenous demand is injected into (see Nakamura, 1999).

DISCUSSION

Our original assumption was that the pattern and magnitude of economy-wide influences of the Russian oil sector would be similar to those of other sectors. This hypothesis was partially accepted and partially rejected. The result of our analysis showed that the extractive sector could induce demand for tradables as much as other tradable sectors. This finding supports our assumption. The expanding oil industry can contribute to fostering domestic tradable sectors through creating demand for them. On the other hand, the oil sector tends to induce more demand for non-tradables than do other tradable sectors. This result disagrees with our hypothesis and confirms that the extractive sector is more likely to cause Dutch disease, and to cause it more strongly than non-extractive tradable sectors. It is difficult to judge how economically significant the difference is. The difference in terms of the absolute value, $A_{NR} - A_{NT}$, is 0.06 in $R01$ and ranged from 0.021 to 0.065 in the OECD cases. Should an expansion of an extractive industry be avoided and an expansion of a non-extractive tradable industry welcomed because of these differences? To answer this question conclusively, we need to extend our model to include price and financial variables. The results can be very different, if we consider price-sensitive non-linear behaviors of economic agents (see Wijnbergen, 1984; Edwards and Ahamed, 1986; Harberger, 1986).

It is true that the Russian oil boom is causing appreciation of the ruble; the appreciation more or less hampers the export of Russian manufactured goods and directs the demand to the imports. Most of Russian manufactured goods are, however, not competitive in the world market with or without the appreciation. If the ruble appreciation does not completely eliminate the demand for domestic manufactured products and the oil boom increases the demand for the Russian manufactured goods, then the boom would turn out to be a help for the Russian manufacturing sector. So far, as Table 3 shows, the shares of imports in the intermediate inputs are small. At the same time, Table 3 indicates that the share of imports in the private consumption is high in Russia. This can be a first symptom of the Russian Dutch disease. We need to have a full-fledged model to consider price and financial effects and identify the outcome of the Russian oil boom.

Despite the limitations, our research showed clearly that the Russian extractive sector could create demand for domestic manufactured goods, hire workers, generate income, and finance economic growth if it was managed well. One more thing should be added: the Russian oil export is not particularly large. The ratios of the Russian exports of fuels and mining to the total merchandise exports of China, Japan, and the USA in 2003 were 19.8 percent, 18.4 percent, and 12 percent respectively (WTO, 2005). It may be an exaggeration if we say that the Russian oil sector generates an extraordinary amount of funds; the amount seems rather small for putting the huge Russian economy back on a sustainable growth path.

REFERENCES

Askari, Hossein, and Mohamed Jaber, "Oil-exporting Countries of the Persian Gulf: What Happened to All That Money," *Journal of Energy Finance and Development,* **4,** 2:185-218, 1999.

Amuzenger, Jahangir, "Oil Wealth: A Very Mixed Blessing," *Foreign Affairs,* **60,** 4:814-835, 1982.

Auty, R., "Industrial Policy Reform in Six Large Newly Industrialising Countries: The Resource Curse Thesis," *World Development,* **22,** 1:11-26, 1994.

Benjamin, Nancy C., Shantayanan Devarajan, and Robert J. Weiner, "The 'Dutch' Disease in a Developing Country: Oil Reserves in Cameroon," *Journal of Development Economics,* **30,** 1:71-92, 1989.

Bosson, Rex, and Bension Varon, *The Mining Industry and the Developing Countries.* New York, NY: Oxford University Press for the World Bank, 1977.

Corden, W. M., "Booming Sector and Dutch Disease Economics: Survey and Consolidation," *Oxford Economic Papers,* **36,** 3:359-380, 1984.

Davis, Graham A., "Learning to Love the Dutch Disease: Evidence from the Mineral Economies," *World Development,* **23,** 10:1765-1779, 1995.

Edwards, Sebastian, and Liaquat Ahamed, eds., *Economic Adjustment and Exchange Rates in Developing Countries.* Chicago, IL: University of Chicago Press, 1986.

Fardmanesh, Mohsen, "Terms of Trade Shocks and Structural Adjustment in a Small Open Economy: Dutch Disease and Oil Price Increases," *Journal of Development Economics,* **34,** 1/2:339-353, 1990.

Fardmanesh, Mohsen, "Dutch Disease Economics and the Oil Syndrome: An Empirical Study," *World Development,* **19,** 6:711-717, 1991.

Gelb, A., "Adjustment to Windfall Gains," in Neary and Wijnbergen (1986a), 54-92, 1986.

Gelb, A., *Oil Windfalls.* New York, NY: Oxford University Press, 1988.

Harberger, A., "Economic Adjustment and the Real Exchange Rate," in Edwards and Ahamed (1986), 371-423.

Investitsii v Rossii, 2003. Moscow, Russia: Goskomstat Rossii, 2003.

Nakamura, Yasushi, "Extractive Industries and 'Dutch Disease,'" *Discussion Paper in Economics,* 99/5, Department of Economics, Heriot-Watt University, 1999.

Nakamura, Yasushi, "The Oil and Gas Industry in the Russian Economy," *Post-Communist Economies,* **16,** 2:153-167, 2004.

Natsional'nye scheta Rossii v 1996–2003 gg. Moscow, Russia: Rosstat, 2004.

Neary, J., and S. van Wijnbergen, eds., *Natural Resources and the Macroeconomy.* Oxford, UK: Blackwell, 1986a.

Neary, J., and S. van Wijnbergen, "Natural Resources and the Macroeconomy: A Theoretical Framework," in Neary and Wijnbergen (1986a), 1986b, 13-45.

OECD, *The OECD Input-Output Database: A Book and Diskettes.* Paris, France: OECD, 1995.

Rodgers, Y., "Empirical Investigation of One OPEC Country's Successful Non-oil Export Performance," *Journal of Development Economics*, **55**:399-420, 1998.

Sachs, J., and A. Warner, "Natural Resource Abundance and Economic Growth," *NBER Working Paper Series*, **5398**, 1995.

Sachs, J., and A. Warner, "The Curse of Natural Resources," *European Economic Review*, **45**:827-838, 2001.

Sistema tablits "Zatraty-Vypusk" Rossii za 2001 god. Moscow, Russia: Rosstat, 2004.

Sitz, A., "Dutch disease," *Jahrbuch fuer Sozialwissenschaft,* **37**, 2:218-245, 1986.

United Nations (UN)**,** *National Accounts Year Book.* New York, NY: UN, various years.

Usui, N., "Policy Adjustments to the Oil Boom and Their Evaluation," *World Development*, **24**, 5:887-900, 1996.

Wijnbergen, S. van, "The 'Dutch Disease'," *The Economic Journal*, **94**:41-55, 1984.

World Trade Organization (WTO), *Statistics Database,* 2005 [http://stat.wto.org./]

Table 1. Structure of the NAM

		1^a	2	3	4	5	6	7	8
1^a	Goods		$●^b$	●		●	●	●	
2	Value added	●						●	
3	Institutional Sectors	●	●					●	
4	Saving			●				●	
5	Inventory Investment				●				
6	GFCF				●				
7	ROW	●	●	●		●			
8	*Total*								

Notes:
[a] The blocks include the following accounts:
  Block 1: labour, capital, net indirect taxes (mixed income in the Russian NAM); *Block 2*: 35 production accounts in the OECD NAM, 25 sector and 25 goods accounts in the Russian NAM; *Block 3*: households, government, corporations, direct tax, social security, property income, other current transfers; *Block 4*: saving, capital transfer, *Block 5*: 1 inventory investment; *Block 6*: 35 by 35 GFCF matrix for Canada and the Netherlands, one GFCF account for Australia, the U.K., and Russia; *Block 7*: an account corresponding to the balance of payments.
[b] The symbol "●" denotes that at least a part of the block has corresponding transactions.

Table 2. Basic Features of the Extractive Sectors and the I-O Tables

Country/Year[a]	A68	A74	A89	C71	C81	C90	N72	N81	N86	U68	U79	U90	R01[g]
Extractive sector[b]													
Gross Output	2.3	3.4	4.3	7.3	3.7	3.8	2.4	3.2	3.2	2.4	3.1	2.1	4.8
Value added	2.7	4.4	4.7	11.0	3.8	4.5	3.9	5.0	5.0	3.7	4.4	2.1	6.6
GFCF[c]	-	-	-	11.3	14.8	7.4	2.2	2.2	2.4	-	-	-	16.6
Export[d]	10.9	18.1	23.4	24.4	11.8	10.8	3.4	5.2	3.8	0.8	6.9	5.1	35.6 / 21.4
I-O tables													
Valuation[e]	B	B	B	P	P	P	P	P	P	P	P	P	B
Industries[f]	33	33	33	35	35	35	33	33	33	35	35	35	24

Notes:
[a] *A*: Australia, *C*: Canada, *N*: the Netherlands, *U*: the United Kingdom, *R*: Russia. The figures denote the years, such as 68 for the year 1968.
[b] The shares (in percent) of the extractive sectors for the OECD countries calculated from the I-O tables at fixed prices. For Russia, they were calculated from the Use table in *Natsional'nye* (2004).
[c] The Russian IOT does not show the gross fixed capital formation (GFCF) by sector. The figure is taken from *Investitsii* (2003, p. 27).
[d] For Russia, the top figure is at purchaser prices; the bottom at basic prices.
[e] *B*: at basic prices, *P*: at producer prices.
[f] The number of production sectors excluding the statistical discrepancy account.
[g] Including the oil refinery sector.
Sources: OECD (1995), *Sistema* (2004).

Table 3. Input and Expenditure Structures, E_{ij} (in percent)[a]

	Extractive sector					Tradable sector					Non-tradable sector				
	L	C	T	N	M	L	C	T	N	M	L	C	T	N	M
A68	28	30	12	12	2	26	15	32	15	10	38	25	12	22	2
A74	23	44	8	11	3	30	11	31	15	11	44	21	10	20	2
A89	15	44	8	16	6	19	19	29	17	12	38	23	10	22	3
C71	21	35	4	27	2	28	12	36	20	23	29	35	11	21	5
C81	14	40	6	39	4	24	13	26	18	16	32	32	8	20	3
C90	16	37	5	34	4	24	13	25	16	20	30	32	8	23	4
N71[b]	84	-	1	3	3	42	-	20	11	28	73	-	9	13	5
N81[b]	85	-	2	7	9	34	-	17	13	32	69	-	9	14	7
N86[b]	83	-	2	7	9	36	-	17	12	33	70	-	9	14	6
U68	54	12	12	7	2	28	10	36	10	11	43	24	15	10	3
U79	20	52	7	14	3	26	10	29	18	17	36	21	13	18	4
U90	17	46	7	10	9	27	13	21	16	33	34	20	10	29	12
Avg[c]	23	39	6	16	5	26	13	27	15	21	36	26	10	19	5
R01b[d]	10	26	6	13	2	20	15	34	9	8	27	27	18	16	4
R01p[d]	8	36	6	9	2	16	18	46	12	9	20	38	21	18	3

	Private consumption				GFCF		
	-	-	T	N	M	-	-	T	N	M
A68			36	59	5			22	68	09
A74			28	66	6			17	71	10
A89			20	63	7			17	64	15
C71			26	67	7			29	41	31
C81			24	67	9			25	47	28
C90			16	72	12			20	42	38
N71			27	51	12			23	41	31
N81			20	67	13			19	39	31
N86			18	69	13			18	34	38
U68			35	60	5			38	56	6
U79			29	59	12			32	52	16
U90			22	63	15			21	59	19
Avg[c]			25	64	10			22	41	33
R01b[d]			19	29	22			11	58	21
R01p[d]			50	29	22			15	64	21

Notes:

[a] The coefficients (E_{ij}) of inputs in terms of percentage of the total output and expenditure ($i=L, C, T, N, M$). L, C, T, N, and M denote compensation for employee, gross operating surplus, tradables, non-tradables, and imports, respectively. The totals are not equal to 100 because of exclusion of the extractive (R) sector products, different treatments of indirect taxes, and rounding errors. For private consumption and GFCF, the tradables include the products of the extractive sector; their shares are less than one percent for all cases.

[b] For the Netherlands, the entries of the column L show the share of total value added. The average is taken excluding the Netherlands figures.

[c] The simple averages of the column entries of the OECD NAMs.

[d] R01b: at basic prices, R01p: at purchaser prices.

Sources: The NAMs compiled by the author.

Table 4. Economy-wide Influences, A_{XY}[a]

| | Influences on Tradable sector (A_{TY})[b] | | | | | | | Influences on Non-tradable sector (A_{NY})[b] | | | | | | |
| | Multipliers | | | | Deviations (%)[c] | | | Multipliers | | | | Deviations (%)[c] | | |
from	P	R	T	N	R	T	N	P	R	T	N	R	T	N
A68	0.18	0.19	0.18	0.18	5.6	0.0	0.0	0.55	0.59	0.53	0.61	7.3	-3.6	10.9
A74	0.14	0.14	0.14	0.14	0.0	0.0	0.0	0.57	0.60	0.55	0.62	5.3	-3.5	8.8
A89	0.09	0.08	0.09	0.08	-11.1	0.0	-11.1	0.42	0.44	0.41	0.46	4.8	-2.4	9.5
C71	0.10	0.10	0.10	0.10	0.0	0.0	0.0	0.35	0.39	0.34	0.39	11.4	-2.9	11.4
C81	0.08	0.08	0.08	0.08	0.0	0.0	0.0	0.30	0.35	0.28	0.33	16.7	-6.7	10.0
C90	0.05	0.05	0.05	0.05	0.0	0.0	0.0	0.27	0.31	0.25	0.30	14.8	-7.4	11.1
N72	0.05	0.05	0.05	0.05	0.0	0.0	0.0	0.17	0.20	0.15	0.21	17.6	-11.8	23.5
N81	0.03	0.03	0.03	0.04	0.0	0.0	33.3	0.14	0.18	0.12	0.18	28.6	-14.3	28.6
N86	0.03	0.03	0.03	0.03	0.0	0.0	0.0	0.15	0.17	0.13	0.18	13.3	-13.3	20.0
U68	0.14	0.15	0.14	0.14	7.1	0.0	0.0	0.33	0.36	0.31	0.36	9.1	-6.1	9.1
U79	0.09	0.09	0.09	0.10	0.0	0.0	11.1	0.27	0.31	0.26	0.31	14.8	-3.7	14.8
U90	0.05	0.04	0.05	0.04	-20.0	0.0	-20.0	0.27	0.28	0.26	0.31	3.7	-3.7	14.8
Avg[d]	0.09	0.09	0.09	0.09	-1.5	0.0	1.1	0.32	0.35	0.30	0.36	12.3	-6.6	14.4
R01	0.15	0.16	0.15	0.15	6.7	0.0	0.0	0.28	0.32	0.26	0.31	14.3	-7.1	10.7

| | Influences on Labor income (A_{LY})[b] | | | | | | | Influences on Capital income (A_{CY})[b] | | | | | | |
| | Multipliers | | | | Deviations (%)[c] | | | Multipliers | | | | Deviations (%)[c] | | |
from	P	R	T	N	R	T	N	P	R	T	N	R	T	N
A68	2.97	3.12	2.85	3.24	5.1	-4.0	9.1	2.14	2.41	2.05	2.34	12.6	-4.2	9.3
A74	3.32	3.33	3.19	3.63	0.3	-3.9	9.3	1.63	2.01	1.55	1.80	23.3	-4.9	10.4
A89	2.04	1.99	1.93	2.30	-2.5	-5.4	12.7	1.71	1.99	1.66	1.84	16.4	-2.9	7.6
C71	2.58	2.68	2.49	2.78	3.9	-3.5	7.8	1.52	1.80	1.41	1.76	18.4	-7.2	15.8
C81	2.18	2.26	2.05	2.45	3.7	-6.0	12.4	1.29	1.72	1.19	1.50	33.3	-7.8	16.3
C90	1.87	1.94	1.76	2.10	3.7	-5.9	12.3	1.14	1.48	1.05	1.35	29.8	-7.9	18.4
N72	1.34	1.68	1.19	1.66	25.4	-11.2	23.9	0.69	0.86	0.61	0.85	24.6	-11.6	23.2
N81	1.05	1.43	0.89	1.37	36.2	-15.2	30.5	0.63	0.86	0.53	0.82	36.5	-15.9	30.2
N86	1.03	1.34	0.89	1.33	30.1	-13.6	29.1	0.74	0.96	0.64	0.95	29.7	-13.5	28.4
U68	2.59	2.97	2.48	2.85	14.7	-4.2	10.0	1.29	1.36	1.22	1.47	5.4	-5.4	14.0
U79	1.85	1.91	1.76	2.09	3.2	-4.9	13.0	0.97	1.45	0.90	1.15	49.5	-7.2	18.6
U90	1.50	1.42	1.43	1.66	-5.3	-4.7	10.7	0.85	1.16	0.80	0.96	36.5	-5.9	12.9
Avg[d]	2.03	2.17	1.91	2.29	9.9	-6.9	15.1	1.22	1.51	1.13	1.40	26.3	-7.9	17.1
R01	1.21	1.34	1.10	1.35	10.7	-9.1	11.6	0.33	0.39	0.30	0.36	18.2	-9.1	9.1

Notes:

[a] Net influences. The initial injection is excluded.

[b] A_{XY}: accounting multipliers ($X=T, N, L, C$; $Y= P, R, T, N$). P, R, T, N, L, and C stand for the average production, the extractive, the average tradable, the average non-tradable, labor income, and capital income, respectively.

[c] *Deviations* are defined as $100*(A_{XY}-A_{XP})/A_{XP}$ ($X=T, N, L, C$; $Y=R, T, N$)

[d] Simple averages of the column entries of the OECD NAMs.

Sources: The NAMs compiled by the author.

Fig. 1. Sensitivity and Power of Dispersion of the Russian Industries

Sensitivity index

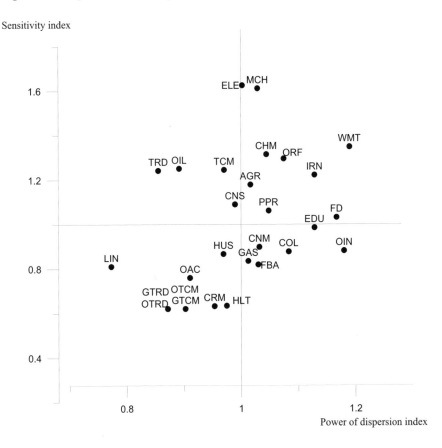

Power of dispersion index

Legends: *ELE*, electricity; *OIL*, oil drilling; *ORF*, oil refinery; *GAS*, gas; *COL*, coal mining; *CRM*, ceramic; *IRN*, iron; *WMT*, nonferrous metal; *CHM*, chemical; *MCH*, metal working and machine building; *PPR*, paper; *CNM*, construction materials; *LIN*, light industry; *FD*, food; *OIN*, other industries; *CNS*, construction; *AGR*, agriculture; *TCM*, transportation and communication; *OTCM* and *GTCM*, transportation and communication related to oil and gas products; *TRD*, trade; *OTRD* and *GTRD*, oil and gas products trade; *OAC*, other activities; *HUS*, housing and communal services; *HLT*, health, education, and culture; *EDU*, higher education and sciences; *FBA*, financial and administrative services.

Notes: The indices were calculated using the products by products section of the calculated Russian NAM. They may differ from those obtained from the original Russian IOT.

Sources: The 2001 Russian NAM compiled by the author.

Chapter 4

Capital Flight from Russia

Akira Uegaki

INTRODUCTION

The transition of the Russian Federation to a market economy has one outstanding characteristic in comparison with other transition countries of Eastern and Central Europe. The difference is that Russia has achieved a tremendous current account surplus in its balance of payments since the collapse of the previous regime, whereas the other transition countries have had deficits in their current accounts for most years since 1989.

The amount of Russia's current account surplus reached 18.4 percent of its GDP in 2000, which is four times that of Japan in the 1980s.[1] From the point of view of accounting, an increase in the current account surplus indicates that an increased amount of a resident's financial resources would remain in a non-resident country. Russia is no exception. In fact, the Russian residents have accumulated much financial resources in non-residents' countries. The problem is that some part of the surplus has fled Russia and it is very unlikely that those resources will soon be returned to Russia.[2] This is the capital flight problem of Russia.

The above situation is different from that of sudden capital flight from a country where a balance of payments crisis is occurring. Continuous capital flight, rather than sudden capital flight, is the focus of this study. Although mainstream economists regard sudden capital flight as a troublesome problem, they less alarmed about continuous capital flight.

---

[1] Calculated using the data of *Natsional'nye* (2002), *IFS*, *RTs*, and the website of the CBR.

[2] As is mentioned later, the financial resources once fled the country has been returning recently.

This is so because, when capital flight is sudden, the market power is overextended, and, as a result, unexpected problems occur. On the contrary, the economists believe that continuous capital flight is the result of a rational economic behavior of the market. The economists assume that capital that has fled a country will return when the economy becomes sound. If it did not, preventing capital from fleeing a country using administrative means would spoil the financial resources.

Nevertheless, the contention held here is that continuous capital flight from Russia has damaged the economy and that government efforts to prevent capital flight would have been worthwhile. The objectives in this chapter are to explain the process of continuous capital flight in Russia and evaluate the attitudes of individuals who have invested abroad.

CURRENCY CRISIS AND CAPITAL FLIGHT

1997 East Asian Currency Crisis and 1998 Russian Currency Crisis

On August 17, 1998, the Russian government made the three following decisions: (1) to widen the exchange rate band (Corridor) to the limits between 6 and 9.5 rubles to one US dollar; (2) to suspend the repayment of the external debt for 90 days; and (3) to suspend sales and refunds of governmental bonds (see, for example, *Rossiiskaia gazeta*, August 17, 1998, p. 1). The first decision affecting the exchange rate led to the acceptance of a floating rate and the depreciation of the ruble. The second and third decisions are, in effect, moratoriums. Therefore, these government decisions are essentially admissions of failure of international macroeconomic policies that have been in effect since 1995.

Although the economic crisis was reported as a shock in Russia and abroad, the details of the crisis remain unclear today. It is noteworthy that the facts indicate that the crisis did not seriously affect the Russian economy. Furthermore, it appears to have triggered the subsequent economic growth. For example, although the monthly index of real consumption of goods and services of Russia declined during the crisis from 107.4 (the average level in 1995 = 100) in August 1998 to 75.5 in February 1999, it rapidly recovered, reaching the 1995 level in December 1999. The trend in the unemployment rate was the same. It was between 11.3 percent and 11.7 percent until the end of August 1998, and it increased to 14.1 percent at the end of February 1999. However, it declined steadily to 12.9 percent at the end of November 1999. In the end, the decrease of the real GDP in 1998 from 1997 was −5.3 percent, and the real GDP increased in 1999 in comparisons with 1998 and 1997.[3]

This was caused by the fact that the Russian banks that were seriously damaged by the crisis had not been doing business without any close connection with the real (that producing goods and services) economy of Russia. Toshihiko Shiobara pointed out that this structure is reflected in the low degree of the financial deepening of the Russian economy (Shiobara, 2004, pp. 191-200).

Table 1 shows the real GDP growth rates of East Asia and Russia before and after their respective crises. In East Asia, the currency crises started in the summer and autumn of 1997, and the effects of the ensuing depression continued through 1998. The Russian crisis started in August 1998, and a rapid recovery has been in effect since the middle of 1999. Therefore, the GDP growth in Russia in 1999 must be regarded differently from that in East Asia in the same year, which reveals that the economic recession in Russia did not last as long as that in East Asia. Light industry in Russia was destroyed before the crisis by the high exchange rate of the ruble; however, it was revived with the currency depreciation after August 1998. In that sense, the crisis can be considered to have triggered the recovery.

Capital Movements of East Asia and Russia during and after the Crises

The amount of capital flight from Russia and countries in East Asia is considered here. Table 2 presents the capital movements of East Asia and Russia taken from the statistics of the balance of payments. The most interesting fact in the table is that four countries in East Asia have experienced capital outflow since 1997 from a category described as "other investment." Net receipts had been recorded in this category until the crisis. Three of four countries, except Malaysia, have lost financial resources from the liabilities entry in the "other investment" category. In other

---

[3] Here, the statistical data cited are from various issues of *Russian Economic Trend: Monthly Update*, October 11, 2000; *Russian Economic Trend*, **11**, 4: 90 (2002); the website of Rosstat [http://www.gks.ru/bgd/free/b01_19/IswPrx.dll/Stg/d000/i000020r.htm].

words, capital that had been invested for bank loans and other lending, including trade credit, was suddenly withdrawn from East Asia after the 1997 crisis. With regard to Indonesia, capital outflow there first occurred under the "other investment" category, followed by portfolio investment and direct investment. These phenomena illustrate that the crisis in Indonesia was more severe than elsewhere.

Russia did not experience such a movement of capital. The structure of the capital movement of Russia was much more complicated. First, securitization was carried out in 1997 to reschedule the former USSR London Club debt (see *Finansovye Izvestiia*, August 11, 1998, p. 8). In this securitization arrangement, the debt of 28.2 billion dollars was supposed to have been repaid and deducted from the liabilities side of the "other investment;" at the same time, an identical amount was supposed to have been borrowed again and added to the liabilities side of the portfolio investment. Therefore, if it had not been for the securitization, the net balance of the portfolio investment in 1997 would have been 17,575 (million dollars), the liabilities of the "other investment," 12,999, and the net balance of the "other investment," −7,634 (see the figures in parentheses).

In 2000, the governmental long-term debt was repaid and re-borrowed under the London Club agreement. The large negative net balance of the portfolio investment in 2000 represents the difference between the repaid amount in the booking record and the re-borrowed amount. Formerly, the difference or the negative figure in the book must have been paid by the Government, but, in reality, the amount was exempted by the decision of the London Club.[4] If it had not been for this arrangement, the net balance of portfolio investment would have been −907 (million dollars) instead of −10,334. As for the negative amount of the liabilities side of the "other investment" in 2000, one half of it was payment of overdue interest on governmental bonds.[5]

Considering all of the conditions reported above, there was no significant capital outflow from the liabilities side of the "other investment. Similarly, the portfolio investment was not withdrawn in a massive scale as in the case of Indonesia, though the new inflows of portfolio invest-

---

[4] The exempted amount is reflected in the positive figure of "capital transfer" in the balance of payments of 2000.

[5] All these facts can be traced in the balance of payments published on the website of the CBR (see the detailed pdf version).

ment decreased.[6] It is true that if we take the residents' capital that fled Russia after the crisis into consideration, the whole amount of capital outflow reached a considerable amount (Uegaki, 2004, Table 5). However, the repatriation of non-residents' capital, a typical phenomenon of countries undergoing a balance of payments crisis, occurred in Russia not as much as in East Asia.

The capital outflow from the liabilities side of the "other investment" for the three years during and after the crisis (from 1997 through 1999 for East Asia and from 1998 through 2000 for Russia) is shown in Figure 1. The capital outflow from the liabilities side of the "other investment" represents the repatriation of the once-invested financial resources of non-residents, such as bank deposits, bank loans, and other trade-related credit. Figure 1 shows that there was a large repatriation from Indonesia, South Korea, and Thailand but not from Russia.[7]

There are obviously many examples of financial and non-financial sector repayments of debt, and there was a large decrease of reserve assets during the crisis in Russia. However, they were overshadowed by other large flows of financial resources. A tremendous amount of the current account surplus before and after the crisis (that is, the years except 1997 and 1998) exists behind the problem.

Table 3 shows the net capital outflow from the private sector of Russia. The net capital outflow means the increase of claims of Russian residents against non-residents or the decrease of claims of non-residents against Russian residents.[8] It is true that the capital outflow from the banking sector increased since the crisis, but the amount was not as big as the outflow from the non-financial enterprises and household sector. The outflow from the household sector remained high even before the crisis.

A similar phenomenon was observed in the movement of the reserve assets of the central bank.[9] The trend of the net change in the reserve as-

---

[6] It must be noted that, in the balances of Eastern Asian countries except South Korea, the active side of the portfolio investment is negligible.

[7] It must be noted that the massive capital outflow from the liabilities side of the "other investment" occurred also in Russia, but it did not occur right after the crisis.

[8] The figure includes the delay of repayments of non-residents' debt to Russian residents.

[9] Precisely, we must use the term "Monetary Authority" which includes other governmental financial institutions. However to use the term "Monetary Authority" is too rigorous and would lead general readers to complicated images of the events. Therefore the

sets of Russia is shown in Table 2. It is noteworthy that the negative figures indicate an increase in the reserve assets of the central bank. Indeed,
the Central Bank of Russia lost more than 5 billion dollars from its reserve
in 1998; however, it is not easy to determine whether the amount was
large enough in comparison to other items in the capital movement. For
example, the average annual capital outflow through a category labeled
"net errors and omission" had been near 9 billion dollars from 1995
through 2000 and always fluctuated, regardless of the crisis. The loss of
the reserve is attributed to the exchange of Russian rubles for US dollars
by non-residents who sold ruble-nominated government bonds. On the
contrary, the net outflow in the category "net errors and omission" may be,
for example, a result of false and illegal statements. Table 2 clearly shows
that the absolute value of the capital movement through the category titled
"net errors and omissions" was far in excess of that in the category titled
"net change of the reserve assets."

The First—and Second—Generation Model of
the Balance of Payments Crisis

Paul Krugman once insisted that the "first-generation model" of the
balance of payments crisis could easily explain the 1998 Russian crisis
(Krugman, 1999, p. 2). According to Krugman, a standard balance of
payments crisis proceeds in the following manner:

A country will have a pegged exchange rate. At that exchange rate, the
government's reserves gradually decline. Then at some point, generally
well before the gradual depletion of reserves would have exhausted
them, there is a sudden speculative attack that rapidly eliminates the last
of the reserves. The government then becomes unable to defend the exchange rate any longer. The government is, sometimes, able to weather
the crisis by calling on some kind of secondary reserves and the capital
that has just flowed out may return and the government reserves may
recover. However the reprieve may only be temporary. Another crisis
may occur, which will oblige the government to call on still further reserves. There may be a whole sequence of temporary speculative at-
tacks and recoveries of confidence before the attempt to maintain the

author uses the Central Bank instead.

exchange rate is finally abandoned (Krugman, 1979, pp. 311-312, with some omissions and modifications).

The first-generation model emphasizes the role of the budgetary deficit, which appears as a result of printing money to finance the currency market intervention to maintain a fixed exchange rate. The model insists that the expectation of inflation leads private investors to withdraw the excess money from circulation by trading it for foreign money at the exchange window. The conclusion is that pegging the rate ultimately becomes impossible if the budget is in deficit regardless of the size of the initial reserves (Krugman, 1979, pp. 315-319).

The first-generation model was criticized by the developers of the second-generation model (Obstfeld, 1986; Obstfeld and Rogoff, 1995), insisting that currency crises can be a self-fulfilling event in which the crisis itself creates the economic pressure under which the government caves in (Obstfeld and Rogoff, 1995, p. 86). According to these researchers, in the financial collapse of Mexico in 1994, there was nothing in the country's underlying fiscal situation to suggest that government was insolvent (ibid., p. 84). They reported that Mexico's pegged rate policy was complemented by a remarkably successful effort to reform other facets of the economy. Nevertheless, the December 1994 currency crisis quickly escalated into a wholesale government liquidity crisis, leaving inflation on the rise and the peso sharply depreciated (ibid., pp. 81-82). This opinion differs from that of the fundamentalist Krugman, who believes that the outcome is a natural result of the rational behavior of investors. Obstfeld and Rogoff hold a slightly more complicated post-modern view. If the dichotomy of sudden capital flight and continuous capital flight is adopted in the present study, the Krugman theory could be interpreted to mean that even a sudden capital flight is rational and unstoppable when the fundamentals within a domestic economy are unsound.

It is noteworthy that the framework of institutions and policies before the Mexican crisis was similar to that of Russia. In both countries, the exchange rate was pegged with a crawling peg system or band system[10], the

---

[10] In Mexico, a pre-announced crawling peg system was first introduced, and then a band system replaced it. On the other hand, in Russia, a band system was first introduced and then replaced by a crawling band system. In the last several months before the crisis, it was returned to a band system once again. The crawling band system is a system in which the highest and lowest levels of the exchange rate are settled at a point in time and another

capital market was deregulated, state-owned enterprises were privatized, and the budget deficit was curtailed before the crisis (Obstfeld and Rogoff, 1995, p. 81; Uegaki, 1999; Uegaki, 2004). Therefore, if the theory of Obstfeld and Rogoff is applied, Russia's crisis cannot be explained by the simple application of the first-generation model.

After the East Asian crisis, Krugman wrote, "I was wrong; Maury Obstfeld was right." He developed a model to explain the East Asian crisis, in which he deliberately considered (1) contagion, (2) the transfer problem, and (3) balance sheet problems of enterprises. Though it is uncertain whether he intended to call it a third-generation model, he persisted in the view that the new model is a reconciliation of the first and the second (Krugman, 1999). Ryuzo Miyao conducted econometric tests of the financial performance of the three East Asian countries and concluded that the crisis in Thailand was mainly caused by fundamental factors, whereas those in Indonesia and South Korea were caused by financial panic (that is, the former can be explained by the first-generation model, and the latter, by the second-generation one) (Miyao, 2003, p. 80).

Although Krugman applied the first-generation model to the 1998 Russian crisis, it is our view that a different model should have been used because of the complex features of the crisis. In particular, it must be noted that the situation of the budget deficit just before the crisis was not as simple as portrayed in his model. In July 1995, the Russian government introduced the ruble-nominated bonds, and the budget deficit was paid by using these bonds rather than by printing money. In addition, the government introduced a new exchange rate identified as "corridor" as a nominal anchor to stop inflation. These policies were strongly recommended by the IMF and foreign advisors. It was true that the reserve assets of the Central Bank of Russia were declining before the crisis; however, it is doubtful that the level of the reserves had reached a risky limit.[11]

However, the second-generation model is not necessarily applicable to the Russian case because of its uniqueness. At least in the following four points, the Russian crisis is not similar to those in East Asia or Mex-

---

set of the highest and lowest levels at a later period (for example 6 months later) is also settled beforehand. The tunnel has a downwards slope, representing the fact that the domestic currency gradually depreciates.

[11] However, it must be admitted that the government bonds market was almost out of control just before the crisis. The first-generation model is applicable to the Russian case in this regard.

ico: (1) The crisis did not strongly affect the real economy, which did not have a close connection with the financial sector; (2) On a net basis, the repatriation of financial resources from Russia did not occur on a massive scale, as it did in East Asia; (3) One-way capital outflow had been occurring through errors and omission before, during, and after the crisis; (4) The current account was in surplus after and before the crisis.

These points would lead to the consideration of the macroeconomic structure of Russia in a wider context.

CONTINUOUS CAPITAL FLIGHT

Current Account Surplus

Among the four points mentioned above, the current account surplus is the basic problem that makes the Russian economy unique. Here, the problem is analyzed from a formal macroeconomic viewpoint.

As is well known, Japan and Russia have had a large current account surplus, whereas the current account of the USA has been continuously in deficit, and the deficit became historically large in the late 1990s and the beginning of the 2000s. As for the budgetary balance, Japan and Russia had large deficits until the late 1990s. Since the turn of the century, however, Russia has dramatically improved its budgetary balance. The USA improved its budgetary balance in the late 1990s, but it is again deteriorating. These phenomena can be formalized by the following equation (see, for example, Krugman and Obstfeld, 1991, pp. 299-303):

$CA = (S^p - I) + (T - G)$,
where I=Investment, G=Government expenditures, CA=Current account, S^p =Private saving, and T=Tax.

This equation means that the current account of a country is the sum of the excess of private savings $[S^p - I]$ and the budgetary surplus $[T - G]$. Table 4 shows how these items (annual average in billions of dollars) are related to each other in several countries.

The table and the equation reveal that there was a large excess of private savings in Russia in the periods of 1995–1997 and 1999–2001. Here, it is noteworthy that the "private savings" includes not only individual household savings but also un-invested company funds, which would go to foreign and domestic financial markets. The amount in excess of 30 billion dollars compares with the German figure in the period of 1995–1997. Although most of the private savings was used to cover the budgetary deficit in the period of 1995–1997, there was still a certain amount left to be invested abroad (including the reserve assets of the central bank). The current account surplus of 7.1 billion dollars equals the amount forced out of the country.

After the currency crisis of 1998, the Russian economy recovered rapidly because of high oil prices and currency depreciation. Interestingly, the economic recovery was not accompanied by a reduction in the current account surplus. On the contrary, it has increased since the crisis. In other words, the excess in private savings is not decreasing.

The continuous existence of an excess in private savings means that the income produced in Russia has not been spent by households or invested in domestic enterprises. Here, an important point is that a considerable part of the income has come from natural resources, including the oil and gas industries. The total share of exports from oil, gas, and petroleum products was 36.6 percent to 50.3 percent from 1995 to 2001 (Tabata, 2002, p. 611). Therefore, the problem lies in the income distribution among the workers in the natural resources sector as well as in the input-output structure surrounding that sector. In addition, the situation is suggestive of a weakness of the Russian financial system that allowed for the transfer of financial resources from capital excess sectors to capital shortage sectors.

On the other hand, from the viewpoint of international balance, the positive current account of a country indicates an outflow of financial assets from the country. Taking this into account, a comparison of current accounts of the listed countries leads to an interesting finding. Russia, together with Japan, funnels financial resources into the world market, whereas the USA, Germany, and Brazil absorb them. Although the amount of financial resources provided by Russia is not as large as that provided by Japan, it was enough to cover the financial shortages of Germany and Brazil in the period of 1999–2001. As the Russian fiscal deficit disappeared in the period of 1999–2001, the excess in private savings was absorbed exclusively by the external financial market, including the American market. It is easy for a country with a healthy financial system, a large middle class, a balanced industrial structure, and a functioning democracy to be a capital provider. The capital once provided for other countries will return with fruits someday. Anyway the situation may be a result of time preference of the people. Is Russia such a country?

Contents of Continuous Capital Flight

As far as the current account records surplus, there must be a certain amount of financial resource outflow to counterbalance the surplus. This is because current account + capital and financial account + errors and omission = 0 by definition according to the 5th version of the balance of payments manual of the IMF. If we bear this in mind, it is interesting that the following three items have been recording minus figures for most of the period from 1992 through 2004. The three items are (1) the increase of foreign cash currency circulating in Russia, (2) export charges not received on time (or import goods and services not received on time), and (3) errors and omission. From an accounting viewpoint, these three items "used" up the current account surplus. Capital flight can be defined in many ways; however, for the purposes of this research, it is defined as the sum of the three items listed above and referred to as "continuous capital flight," when applied to Russia. This is so because the minus figures for the three items mean that the financial resources have fled the country through a route that would not put the resources back into the country in the near future.

The idea that the continuous capital flight includes the increase of foreign cash currency circulating in Russia might be disturbing to many. Some researchers would insist that such circulation should not be included in the category of capital flight because it remains in the country. According to the author's view, however, foreign cash currency had the same effect as capital flight because it was hoarded and hidden rather than taken to financial institutions. The main point is that the foreign cash currency was not a resource for investments. Furthermore, foreign cash could be easily smuggled out of the country.[12] It is also noteworthy that foreign currency escaped tax collection.

Export charges not received on time take the following form in Russia. A Russian national delivers goods to a foreign country, and this activity is registered as an export in the statistics for the balance of payments.

---

[12] Of course, some resources may be secretly repatriated.

This Russian national, however, presents a statement to officials claiming that export charges have not been obtained for some reason. In this case, a certain amount of export charges not received on time is recorded in the debit side of the balance of payments (resulting in the negative figures in the balance). Sometimes, this is a false statement to the effect that secret payments are made to the exporter's account of a bank, for example, in Bahamas. This is a typical case of illegal capital flight. It is also true that there might only be a technical problem that has caused the delay in the payment. But negative figures in the item of export charges not received on time caused by a technical problem would be counterbalanced by positive figures in the same item in the long run. In reality, export charges not received on time have recorded a considerable negative amount every year since 1994 through 2004, which reveals that artificial and continuous capital flight has been occurring.

A case of errors and omission is recorded when a transaction is made and registered in the credit (debit) side of the balance of payments; and at the same time, the counter transaction, which must be registered in the debit (credit) side, is not reported to the statistical office.[13] What is important is that a case of errors and omission can appear in positive as well as in negative figures. Nevertheless, in the Russian case, the balance of payments has recorded cases of errors and omission in large negative figures every year from 1992 through 2004 (except 1994). These incidents rarely occurred in other transition countries, and, when they did, they reveal artificial and sometimes illegal capital flight.

Not all the transactions reflected in these three items are necessarily illegal, but some of them are. We have no exact information to decide which is legal and which is not. Therefore we call them "gray" transaction. Regardless of whether the transactions are legal or illegal, the sum represents a reduction in the level of welfare of the Russian people. In this sense, this continuous capital flight harmed the Russian economy.

This opinion is based on an evaluation of the macro-economy of Russia. As reported above, there has been an excess of private savings in Russia. However, this excess is not a reflection of an affluent society, as it is in Japan, which has large private savings. Russia is still a poor country,

---

[13] In 1992, "export charges not received on time" could not be identified separately. Thus, they were combined with "errors and omission" whenever they occurred (see Table 5).

and there are many consumer goods that people would buy if they had money. It is well known that the social infrastructure is poor in Russia. For example, per capita calorie consumption in Russia is lower than that in Poland and Romania and 80 percent of that of the people in Portugal (2002, *RSM*, 2004, p. 97).[14] The so-called Engel's coefficient (share of the expenditure for food, beverages, and tobacco in the total expenditure of a household) is 38.7 percent, which is 1.5 times higher than that of Mexico (1999, *RSM*, 2004, p. 107). The number of personal computers per 1,000 inhabitants is 89 in Russia, whereas it is 431 in Germany, 106 in Poland, and 82 in Mexico (2002, *RSM*, 2004, pp. 255-256). Therefore, there is potential demand in Russia, but it has not been realized because of the uneven distribution of income. In such a situation, if the income earned by the oil and gas industries had not been kept abroad but, rather, had been spent in the domestic market, it would have resulted in economic circulation in Russia. In addition, only a fair tax payment would have contributed to an earlier reinforcement of the infrastructure.[15]

Placing a Figure on Continuous Capital Flight

The three items of capital flight, as they are defined in this research, are shown in Table 5; the sum increased until 1997, then stagnated, and later grew again. According to the table, from 1992 to 1998, the capital flight exceeded the current account surplus every year (except 1993) or recorded a considerable amount even when the current account itself was negative. It is surprising that the amount more than the trade surplus of goods and services[16] had fled the country via the three routes.[17] The shortage was covered by capital inflow. In fact, the net receipts of the real aggregate net capital transfer,[18] which can be called "legal capital inflow", had been recorded until 1998 (Uegaki, 2004, pp. 39-43). Therefore the

---

[14] Of course the high calorie consumption does not necessarily mean a highly civilized life, according to modern nutritional sciences.

[15] Here, the problem that the Russian Government is a very inefficient player in the field of economics is ignored.

[16] Other items in the current account ("current transfer" and "net receipts and payment of interests, dividends, and wages"), are minus or negligible.

[17] A causal relationship of time is not necessarily assumed here.

[18] It includes disbursements of banks and other organizations, long-term trade credit, portfolio investment, and direct investment, considering interest receipts and payments.

sum of the trade surplus and the legal capital inflow streamed out through the gray routes (see Figure 4).

After the 1998 currency crisis, the structure has changed, and the fled capital was covered thoroughly by the current account surplus, though the total amount of the capital flight did not decrease. Here, the tremendous current account surplus corresponds with the gray and legal outflow of capital (see Figure 4). The 1998 currency crisis represents a divide in the history of the international financing of Russia. It is true that the depreciation of the ruble after the crisis made importing unfavorable, which promoted the current account surplus and triggered the rebirth of light industry in Russia. However, this effect did not last long. The real impact on the Russian economy was given by the rapid rise in oil prices, which had nothing to do with the 1998 crisis. The rise in oil prices lifted the current account surplus to a historically high level. It stimulated domestic investments in several sectors[19] of the economy. It has also helped to resolve the fiscal deficit problem, which, in turn, encouraged the reliance on the ruble in the domestic market. The latter point is reflected in the fact that the foreign cash currency circulating in Russia began to rapidly decrease (the positive figures in the table indicate a reduction) in 2003. Nevertheless, the continuous capital flight remained significant until recently. The 1998 currency crisis did not leave any effect in the trend of the continuous capital flight.

The Reasons for the Continuous Capital Flight

The correspondence between the movement of continuous capital flight (the total of the three items) and that of the current account is shown in Figure 2. The larger the current account surplus, the more the capital flight emerges. As each of the three items varied in the degree of correspondence with the current account, the results indicate that the three items are complementary. However, several exceptional periods must not be ignored. For example, from the second to the third quarter of 1995, the capital flight grew, while the current account decreased. In addition, from the second to the fourth quarter of 1997, the capital flight grew, while the current account was in deficit. On the contrary, from the third to the fourth quarter of 1998, although the current account grew, the capital flight de-

---

[19] The oil- and gas-related industries.

creased. In all these periods, irregular movement of the "foreign cash currency circulating in Russia" was evident and caused irregular movement of the total amount of the capital flight (see Figure 3). The irregular movement of the "foreign cash currency circulating in Russia" was brought about by attitudes of the Russian residents in the foreign exchange market and the Central Bank's intervention into the market to keep the exchange rate level, right before, during, and immediately after the Corridor period.

In spite of the above-mentioned exceptional examples, the general tendency towards more capital flight in relation to more current accounts is undeniable, as shown in Figure 2. Here, the following assumption can be made. A certain percentage of financial resources gained from trade activities (the main source of the current account surplus) was not exchanged into rubles but often remained in foreign currency (dollars) or was transferred to foreign countries via a gray route. In particular, before the 1998 currency crisis, the source of the capital that left the country was not only the trade surplus, but also the capital that had once entered Russia legally (see Figure 4).

This stems from the fact that any Russian national or legal resident of Russia can hold foreign currency in cash or in the form of bank deposits in Russia. To keep property in the form of foreign currency means to have property with an internationally recognized value. In such a case, the property can be easily moved from one place to another. This is a result of the liberalization of the foreign monetary and financial system at the beginning of the reform. In Russia, liberalization policies for capital movement were carried out to introduce capital from abroad. Such policies were effective, especially in 1996 and 1997. This is, however, one side of the coin. The other side is a simple liberalization policy of foreign currency, which promoted the outflow of financial resources from the beginning (see Uegaki, 2004, pp. 24-33).

The Rationality of the Behavior

Focusing on the behavior of the economic players of Russia, capital flight is not necessarily an evil thing. Some of the investors, after examining the world financial situation, might have decided to maintain their investments abroad while knowingly accepting the danger of incurring the accusation of the tax offices or other government authorities. It is a persuasive argument that to invest money into the Russian land is risky business and that to keep money in safe places abroad is profitable even for Russians as a whole in the long run. In fact, we see much investment coming back from the so-called "tax haven" such as Bahamas, Luxembourg and others recently. This suggests that the capital that once left Russia is now returning. Whether these investments and decisions were rational from an economic point of view should be examined.

To investigate this problem, it will be necessary to study the trend of the exchange rate of the ruble because capital flight by the Russians reflects their will to have financial resources in foreign currency. The exchange rate should work as a key determinant. The relationship between capital flight and the real exchange rate is shown in Figure 5. The real exchange rates are plotted as dollars per ruble[20] indexed by the consumer price index (the value on the first quarter of 1994 is 100). The rising trend indicates an appreciation of the ruble, and the decline indicates depreciation.

The relationship between the two lines raises an interesting problem. In the "Corridor" period (from the second quarter of 1995 through the second quarter of 1998), the real exchange rate of the ruble was kept at a high level, whereas capital flight occurred on a massive scale. In this period, the Russian Government sought to attract foreign capital by the route of ruble-nominated government bonds and the system of the Corridor. In fact, many foreign investors agreed with this policy and invested in Russian bonds. In contrast to the foreign investors, some of the residents preferred to leave their financial resources abroad through the gray route under the high ruble exchange rate. If they had been rational players, they might have accumulated money abroad (after exchanging rubles for dollars or euros) expecting the collapse of the Corridor system and the decline of the ruble exchange rate. As it turned out that the Corridor system in fact collapsed, those who left their money abroad made a good decision. However, whether they were rational in the sense of Krugman's first-generation model is not easy to answer because the inflation subsided in 1997.[21]

---

[20] Period average.

[21] If we take the mechanism of trades between residents and non-residents into account, the situation becomes more complicated. Especially, if we consider derivative trad-

The time since the first quarter of 2000 also needs to be examined because capital flight increased again as the exchange rate began to increase. It is not rational to invest in foreign currencies when the domestic currency is expected to appreciate. In this period, although the Government had been intervening in the market, its purpose was to prevent the domestic currency from appreciating. To outwit the Government as in the crisis period, the Russian investor should have invested in the ruble. Curiously, some Russians did not.

Besides the exchange rate, the interest rate is an important factor for investors who have alternate ways of investment in domestic and foreign markets. The relationships between capital flight and the risk premium are shown in Figures 6 and 7. Here, the risk premium equals the "domestic real interest rate per annum in Russia" (nominal interest rate indexed by producer inflation rate) minus "LIBOR" (per annum, in euros). Therefore, if the risk premium is higher than 0, there is a chance to earn additional interest by investing in the Russian domestic market rather than in the foreign market.[22] Of course, the risk premium also indicates a high risk of failure and may be a factor discouraging someone from investing. Nevertheless, many foreign investors invested in the Russian high-interest market, accepting the high risk before the crisis. It is natural to think that resident-investors would invest in the Russian market rather than leave their financial resources abroad when the risk premium was high.

Figure 6 indicates that some of the residents acted not as expected by "theory." In the first quarter of 1995, the risk premium jumped from the previous quarter; whereas the capital flight increased (the increase of the capital flight is plotted in minus figures according to the system of balance of payments). In the second quarter of 1995, the risk premium decreased to 0, and investing in the domestic market became less profitable. However, the outflow of capital via the gray route stagnated. In the third quarter of 1995, the risk premium increased again, but more than 4 billion dollars left the country via the gray route. From the second quarter of 1996 to the fourth quarter of 1997, the Russian Government pursued ac-

---

ing, it is not easy to identify who has rational behavior. Alexei Medvedev found that residents and non-residents behaved differently in the governmental bonds market during the crisis (November 1997–August 1998) and that non-residents were more sensitive to negative external news and some domestic news. He asserted that non-residents strongly contributed to the negative developments (Medvedev, 2001, p. 19).

[22] Here, we simply assume that the inflation rate of foreign countries is 0.

tive measures to introduce foreign capital in the domestic governmental bond market under the relatively low rate of inflation. Therefore, the risk premium was stable at 10 to 25 percent. However, a considerable amount of capital left the countries during this period.[23]

Figure 7 shows that the very unprofitable risk premium (in fact, minus rate) caused much capital flight in the period from the fourth quarter of 1998 through the second quarter of 2004.[24] Generally speaking, those who invested abroad via gray routes after the crisis were rational players in the international financial market. If the fluctuations are carefully observed, however, their rationality is doubtful. There are many cases in which the upward (downward) trend of the risk premium is synchronized with the increasing (decreasing) trend of capital flight.

The latter phenomena are also observed in Figure 6. Those who participated in capital flight may be rational economic players in some cases, but their actions did not correspond with subtle changes of the economic environment.

As shown in Figures 5, 6, and 7, it is clear that a certain amount of capital flight occurred when the source for it, current account surplus and in-coming capital, existed. These findings suggest the pessimism that some of the Russians felt for their economy. In the end, when they obtained financial resources in foreign currency, they would prefer to keep some of it in foreign currency regardless of economic indicators, such as interest and exchange rates. It is interesting to find this pessimistic attitude even in the prosperous economy driven by high oil prices since 2002.

CONCLUDING REMARKS

The financial crisis of 1998 did not cause sudden large capital repatriation from Russia, whereas that occurred in East Asia in the 1997 crisis. The crisis in Russia did not affect the real sectors of the economy seriously, which is, again, not similar to the situation in East Asia. Therefore, neither the first-generation model of balance of payments crisis nor the

---

[23] Some would assert that more capital might have flown into the country by the legal route. However, this was not the case (see Figure 4).

[24] As for the third quarter of 1998, it was a period of chaos after the crisis when the nominal interest rate was still extremely high but the inflation rate was not as high as in the subsequent three periods. Therefore, the real interest rate was still high.

second-generation model is applicable to the Russian case. If we investigate the financial flows of Russia from a longer-term perspective, however, we find that capital flight occurred on a massive scale almost every year after the collapse of the USSR. This capital flight was continuous. According to the definition presented here, this continuous capital flight occurred through three routes, which are not necessarily illegal but are potentially harmful to the welfare of the Russians, at least, over the short term. It is debatable whether those who participated in capital flight were acting rationally; however, it is clear that they were not expert financial strategists with the skill to respond to subtle financial indicators. The activities of the Russians indicate pessimism towards their economy; regardless of the degree of prosperity of their economy, they will always keep some of their earned foreign currency abroad.

Our task hereafter is to investigate the relationship between capital flight and the domestic economy in a more numerical way. For example, the problem of whether capital flight caused a reduction in GDP, maybe with a time lag, has not been answered in this paper. The problem of the relationship between capital flight and tax revenue is also a difficult issue. Rigorous econometric tests could be used to answer these questions.

REFERENCES

Borisov, S. M., "Statistika svidetel'stvuet: strana zhivet vzaimy (zametki o platezhnom balans Rossii)," *Den'gi i kredit,* 12:56-69, 1997.

Central Bank of Russia (CBR) = Bank Rossii, Website [http://www.cbr.ru/ statistics].

International Financial Statistics (IFS). Washington, DC: International Monetary Fund, annual and monthly.

Krugman, Paul, "A Model of Balance of Payments Crises," *Journal of Money, Credit, and Banking,* **11,** 3:311-325, 1979.

Krugman, Paul, "Balance Sheets, the Transfer Problem, and Financial Crisis," cited from *Krugman's Website,* 1999 [http://web.mit.edu/ Krugman/www/FLOOD.pdf].

Krugman, Paul, and Maurice Obstfeld, *International Economics, Theory, and Policy*, 2nd ed. New York, NY: Harper Collins Publishers, 1991.

Medvedev, Alexei, *International Investors, Contagion, and the Russian Crisis*, BOFIT Discussion Papers, No.6. Helsinki, Finland: Bank of Finland, Institute of Economies of Transition [BOFIT], 2001 [http://www.bof.fi/bofit/eng/6dp/abs/pdf/dp0601.pdf].

Miyao, Ryuzo, "Origins of the Asian Crisis: Tests of External Borrowing Constraints," in Shinji Takagi, ed., *Currency Crisis and Capital Flight: Reexamination of Asian Currency Crisis.* Tokyo, Japan: Keizai Shinpo Sha, 2003, 61-83 (in Japanese).

Natsional'nye scheta Rossii v 1994–2001 gg. Moscow, Russia: Goskomstat Rossii, 2002.

Obstfeld, M., "Rational and Self-fulfilling Balance-of-Payments Crises," *American Economic Review,* **76**:72-81, 1986.

Obstfeld, M., and K. Rogoff, "Mirage of Fixed Exchange Rates," *Journal of Economic Perspectives,* **9**, 4:73-96, 1995.

Rossiia i strany mira (RSM). Moscow, Russia: Rosstat, 2004.

Rossiia v tsifrakh (RTs). Moscow, Russia: Rosstat, various years.

Rosstat (Federal'naia sluzhba gosudarstvennoi statistiki; formerly Goskomstat), Website [http://www.gks.ru/wps/portal].

Shiobara, Toshihiko, *Economic Structure of Present Russia.* Tokyo, Japan: Keio University Press, 2004 (in Japanese).

Tabata, Shinichiro, "Russian Revenues from Oil and Gas Exports: Flow and Taxation," *Eurasian Geography and Economics,* **43**, 8:610-627, 2002.

Uegaki, Akira, "Corridor and the GKO/OFZ: Macro-economy and International Finance of the Russian Federation," *Economic Review of Seinan Gakuin University,* **33**, 4:107-154, 1999 (in Japanese).

Uegaki, Akira, "Russia as a Newcomer to the International Financial Market, 1992–2000," *Acta Slavica Iaponica*, **21**:23-46, 2004 [http://src-h.slav.hokudai.ac.jp/index-e.html].

Table 1. Real GDP Annual Growth (in percent)

	1996	1997	1998	1999
Indonesia	8.0	4.5	-13.0	0.3
Malaysia	10.0	7.3	-7.4	5.6
Korea	6.8	5.0	-6.7	10.7
Thailand	5.9	-1.7	-10.2	4.2
Russia	-3.6	1.4	-5.3	6.3

Sources: Rosstat website [Russia] and *World Economic Outlook*, October 2000 [other countries].

Table 2. Capital Movement of Eastern Asia and Russia (millions of dollars)

Country	Item	1994	1995	1996	1997	1998	1999	2000
Indonesia	Current Account	-2792	-6431	-7663	-4889	4096	5785	7985
	Direct Investment, Net Balance	1500	3743	5594	4499	-356	-2745	-4550
	Portfolio Investment, Net Balance	3877	4100	5005	-2632	-1878	-1792	-1909
	Other Investment Assets					*-44*	*-72*	*-150*
	Other Investment Liabilities	*-1538*	*2416*	*248*	*-2470*	*-7360*	*-1332*	*-1287*
	Other Investment, Net Balance	-1538	2416	248	-2470	-7404	-1404	-1437
	Net Change of Reserve Assets	-784	-1573	-4503	5113	-2090	-3342	-4851
	Net Errors and Omissions	-263	-2255	1319	-2645	1849	2128	3637
Malaysia	Current Account	-4520	-8644	-4462	-5935	9529	12604	8488
	Direct Investment, Net Balance	4342	4178	5078	5137	2163	2473	1762
	Portfolio Investment, Net Balance	-1649	-436	-268	-248	283	-1025	-2532
	Other Investment Assets	*504*	*1015*	*4134*	*-4604*	*-5269*	*-7936*	*-5565*
	Other Investment Liabilities	*-1909*	*2885*	*533*	*1912*	*272*		
	Other Investment, Net Balance	-1405	3900	4667	-2692	-4997	-7936	-5565
	Net Change of Reserve Assets	3160	1763	-2513	3875	-10018	-4712	1009
	Net Errors and Omissions	154	-762	-2502	-137	3039	-1273	-3221
South Korea	Current Account	-3867	-8507	-23006	-8167	40365	24477	12241
	Direct Investment, Net Balance	-1651	-1776	-2345	-1605	672	5135	4284
	Portfolio Investment, Net Balance	6232	11712	15101	14384	-1224	9190	12177
	Other Investment Assets	*-7369*	*-13991*	*-13487*	*-13568*	*6693*	*-2606*	*-2289*
	Other Investment Liabilities	*13632*	*21450*	*24571*	*-8317*	*-13868*	*1502*	*-1268*
	Other Investment, Net Balance	6263	7459	11084	-21885	-7175	-1104	-3557
	Net Change of Reserve Assets	-4614	-7039	-1416	11875	-30968	-22989	-23790
	Net Errors and Omissions	-1816	-1240	1095	-5010	-6225	-3536	-561

	1	2	3	4	5	6	7
Thailand							
Current Account	-8085	-13554	-14691	-3021	14243	12428	9313
Direct Investment, Net Balance	873	1182	1405	3315	7185	5757	3389
Portfolio Investment, Net Balance	2481	4081	3544	4528	356	-111	-706
Other Investment Assets	*-1027*	*-2738*	*2661*	*-2555*	*-3407*	*-1755*	*-2203*
Other Investment Liabilities	*9839*	*19383*	*11876*	*-17343*	*-18243*	*-14964*	*-10914*
Other Investment, Net Balance	8812	16645	14537	-19898	-21650	-16719	-13117
Net Change of Reserve Assets	-4169	-7159	-2167	9900	-1433	-4556	1608
Net Errors and Omissions	87	-1196	-2627	-3173	-2828	33	-685
Russia							
Current Account	7844	6963	10847	-80	219	24616	46839
Direct Investment, Net Balance	408	1460	1656	1681	1492	1102	-463
Portfolio Investment, Net Balance	21	-2444	4410	45775 (17575)	8618	-946	-10334 (-907)
Other Investment Assets	*-19556*	*-154*	*-27663*	*-20633*	*-14463*	*-13219*	*-17659*
Other Investment Liabilities	*6968*	*14021*	*16080*	*-15201 (12999)*	*9029*	*-889*	*-4172*
Other Investment, Net Balance	-12588	13867	-11584	-35834 (-7634)	-5434	-14108	-21831
Net Change of Reserve Assets	1896	-10386	2841	-1936	5305	-1778	-16010
Net Errors and Omissions	9	-9113	-7708	-8808	-9817	-8558	-9156

Sources: Website of the CBR [Russia] and *IFS* [other countries].

Table 3. Net Capital Outflow from the Private Sector of Russia (billions of dollars)

	1994	1995	1996	1997	1998	1999	2000
From the Banking Sector	-2.0	6.8	1.3	7.6	-6.0	-4.3	-2.1
From Non-Financial Enterprises and the Household Sector	-12.4	-10.7	-25.1	-25.9	-15.7	-16.5	-22.8
Total	-14.4	-3.9	-23.8	-18.2	-21.7	-20.8	-24.8

Note: Minus figures mean capital outflow from Russia.
Sources: Website of the CBR.

Table 4. Annual Average of Macro Statistics (billions of dollars)

		CA = Current account	T - G = Budgetary surplus[a]	S^P-I = CA - (T - G) = Excess of private savings[g]
Russia	1995-1997	7.1	-23.3[b]	30.4
	1999-2001	35.4	4.4[b]	31.0
USA	1995-1997	-123.6	-86.5	-37.1
	1999-2001	-395.5	167.9	-563.4
Japan	1995-1997	90.1	-192.8[c]	282.9
	1999-2001	104.3	-291.5[c]	395.8
Germany	1995-1997	-9.9	-40.2[d]	30.3
	1999-2001	-10.9	-7.4[d]	-3.5
Brazil	1995-1997	-24.0	-59.1[e]	35.1
	1999-2001	-24.1	-4.7[f]	-19.4
South Korea	1995-1997	-10.4	2.2	-12.6
	1999-2001	9.3	-3.2	12.5

Notes:

[a] Converted from each national currency to a US dollar value by exchange rates [yearly average] quoted in *IFS*.

[b] Excluding Social Security funds and extra-budgetary spending.

[c] Calculated from newly issued state bonds in every fiscal year [April to March].

[d] Including special spending for the unification.

[e] Figure in 1997.

[f] Average of 1999 and 2000.

[g] Calculated as a residual [CA - (T - G)] rather than calculated from indigenous sources.

Sources:

Calculated by the author using the data of *IFS*, No.2, 2002 and Data of the Economic Planning Agency of Japan [for Japan's budgetary surplus].

Table 5. Continuous Capital Flight (millions of dollars)[a]

	1992	1993	1994	1995	1996	1997	1998	1999	2000	2001	2002	2003	2004
1. Increase of foreign cash currency circulating in Russia[b]	-1489	-2751	-5740	134	-8740	-13444	824	921	-888	-1123	-1080	5911	3323
2. Export charges not received on time (or import goods and services not received on time)	—	-3600	-4085	-5239	-10119	-11591	-7959	-5051	-5293	-6388	-12244	-15435	-25903
3. Errors and omission	-386	-1167	9	-9113	-7708	-8808	-9817	-8558	-9156	-9481	-6501	-7199	-8385
Continuous capital flight (1 + 2 + 3)	-1875	-7518	-9816	-14218	-26567	-33843	-16952	-12688	-15337	-16992	-19825	-16723	-30965
Current account	-69	9013	7844	6963	10847	-80	219	24616	46839	33935	29116	35905	59935

Notes:

[a] Positive and negative figures are used according to the system of balance of payments. Therefore, the negative figures mean the outflow of financial resources from Russia.

[b] Minus signs mean increase.

Fig. 1. Outflow of Capital from the Liabilities Side of the Other Investment in the Three Years during and after the Crisis

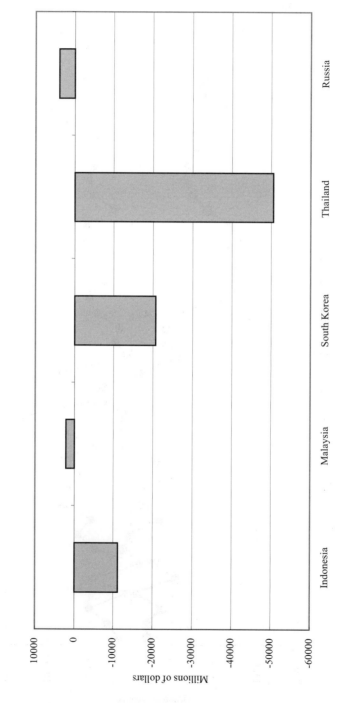

Sources: Same as Table 2

Fig. 2. Capital Flight and Current Account

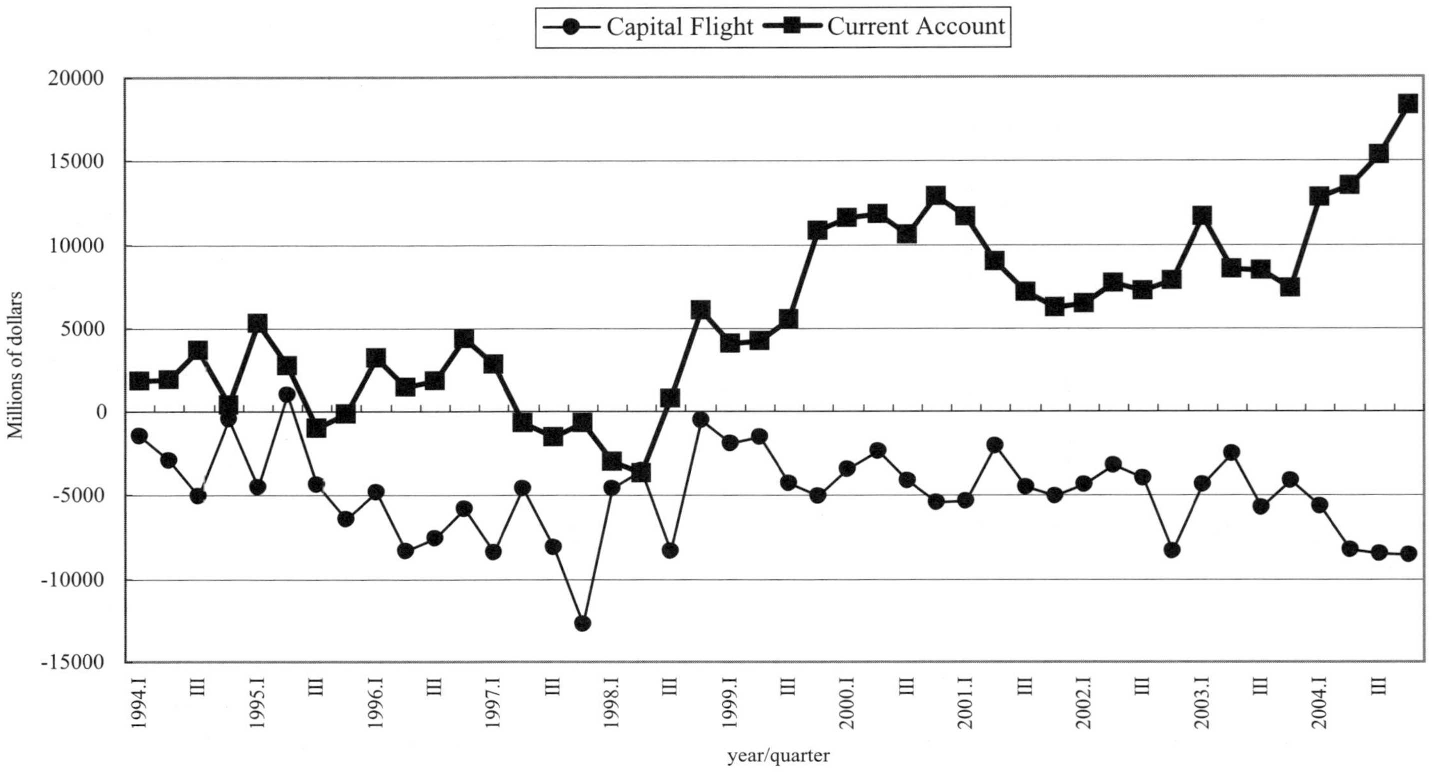

- Capital Flight - Current Account

year/quarter

Millions of dollars

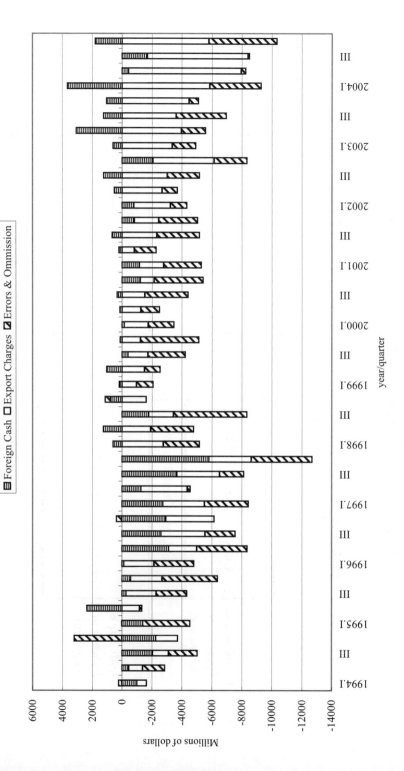

Fig. 3. Three Routes of Capital Flight

Fig. 4. Structure of the Russian International Financing

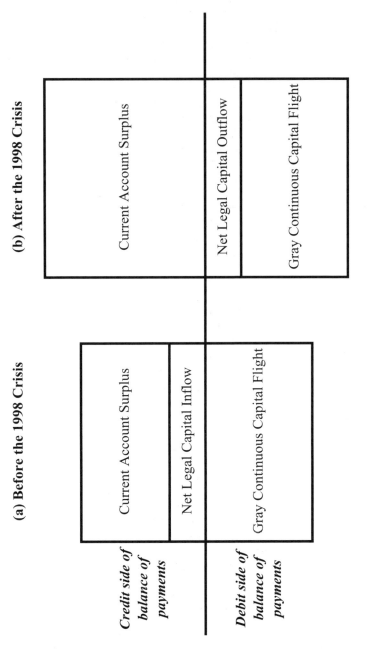

(a) Before the 1998 Crisis

(b) After the 1998 Crisis

Credit side of balance of payments

Current Account Surplus

Net Legal Capital Inflow

Current Account Surplus

Net Legal Capital Outflow

Debit side of balance of payments

Gray Continuous Capital Flight

Gray Continuous Capital Flight

Note: Reserve assets, short-term trade credit, and other miscellaneous credit are omitted from the figure.

Sources: The author's original figure.

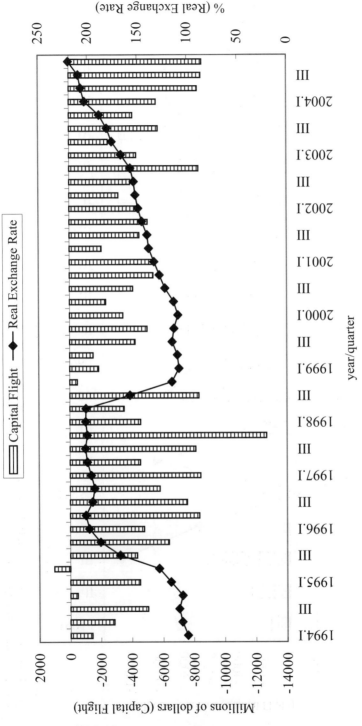

Fig. 5. Real Exchange Rate and Capital Flight

Note: Real Exchange Rate: US dollar per ruble indexed by consumer price index (1994/1 = 100).
Sources: Website of the CBR and *IFS*.

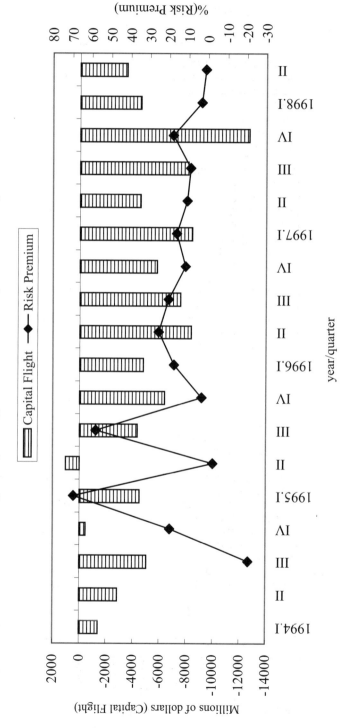

Fig. 6. Capital Flight and Risk Premium (before the Crisis)

Note: Risk Premium = Real Interest Rate in Russia − LIBOR (per annum).
Sources: Website of the CBR and *IFS*.

Fig. 7. Capital Flight and Risk Premium (after the Crisis)

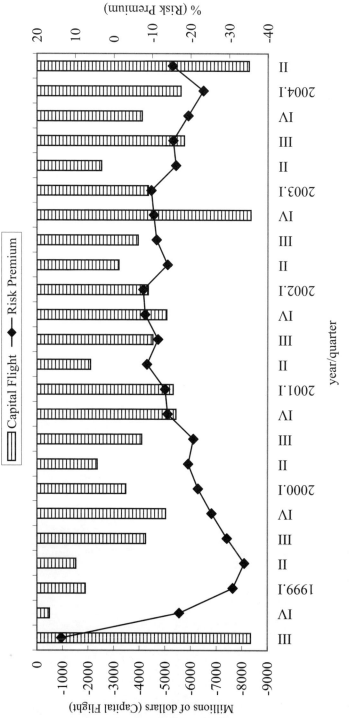

Note: Risk Premium = Real Interest Rate in Russia − LIBOR (per annum).
Sources: Website of the CBR and *IFS*.

Chapter 5

Oversights in Russia's Corporate Governance: The Case of the Oil and Gas Industry[*]

Toshihiko Shiobara

INTRODUCTION

Russia's corporate governance has been discussed widely in academic articles.[1] Nevertheless, very few of these papers effectively analyzed its "reality," which is interpreted here as real activities and real incidences. As Berglöf and Thadden (1999, p. 3) rightly indicate, corporate governance, defined as the mechanisms related to the decision-making process of firms, is likely to matter more in certain contexts or certain phases of economic development than in others. The process of determining when, why, and how much corporate governance matters necessitates asking empirical questions. Therefore, observation and an appreciation of the "reality" in each country is essential. In case of Russia, if we are not focusing on corporate groups in Russia, understanding the present "reality" in the Russian micro-economy is nearly impossible, because the concentration of sales, employment, and ownership by corporate groups has been developed. Therefore, in Russia, analysis of "corporate groups' governance" takes clear precedence over that of "corporate governance," and

[*] This study was financially supported by a grant-in-aid from the Ministry of Education, Science, Culture, Sports and Technology of Japan (Grant # 2005-06).
[1] See for example Afanas'ev, et al. (1997), *Corporate Governance in Russia* (2003), Dolgopiatova (2002, 2003), Iwasaki (2003), Judge and Naoumova (2004), Kostikov (2003), Kuznetsova and Kuznetsov (1999), Radygin (1999, 2002), Roberts (2004), and OECD (2002).

this is a key point for making geographical argument in relation to the oil and gas industry.

Another "reality," which most economists have left unnoticed, is that the security department of firms in Russia plays a big role in conducting economic activities. I have not yet found any critical discussion in which this point is openly addressed to make a case for corporate governance. It is well known that the head of the fourth section of domestic economic security in Yukos was arrested. This means that there are at least four sections related to economic security. Economic security sections are concerned not only with guarding firms, but also with collecting the data of rivals and dealing with security authorities. A number of large firms and banks have similar departments. Many former KGB staff and the police have been employed to work in these departments. Therefore, in Russia, the concept of economic security plays a great role in corporate activities. The importance of economic security has been increased as the role of information has grown. In other words, "competitive intelligence," defined as "a systematic and ethical program for gathering, analyzing and managing information that can affect a company's plans, decisions, and operations," in Russia should be the focus of future efforts (Nasheri, 2005, p. 73).

This chapter, first, will provide an overall view of corporate governance in Russia. Then, the first oversight, the "corporate groups' governance," will be analyzed. This argument leads us to conclude that Russian corporate groups' governance has been less developed, and, therefore, the legislation of applicable laws concerning holding companies, affiliated companies, and transfer pricing is very important. Lastly, some concluding remarks will discuss future issues in relation to corporate governance.

OVERALL VIEW OF CORPORATE GOVERNANCE

The importance of corporate governance depends on the level of economic development and the extent of capitalist economy. It is indicated that factors like investment opportunity and financial system are generally neglected, while the significance of financing from the stock market is overestimated, because corporate governance has been advocated primarily by developed countries (Berglöf and Thadden, 1999, pp. 5-6). Berglöf and Thadden define corporate governance as the set of mechanisms that translate signals from product markets and input markets into firm behavior (ibid., p. 11). This definition focuses on two components: the signals generated outside the firm and the control structures inside the firm to execute decisions based on these signals. These signals are related to competition in markets and the control structure to execute decisions, which is related to corporate integration. The pressure from outsiders is only a part of mechanisms related to corporate governance. Employees, suppliers, competitors, and the Government are also involved in corporate governance. In transition economies, soft budget constraints, "strong insiders," and the "absence" of outsiders should be emphasized when analyzing corporate governance (ibid., pp. 19-21).

In this chapter, corporate governance can be defined as "mechanisms related to the decision-making processes of firms." This definition attempts to draw attention to how, why, and when firms are organized and operated. This view enables us to construct arguments about corporate governance comprehensively. However, in Russia, the meaning of corporate governance is rather narrow, which is apparently made clear when its Russian equivalent, *korporativnoe upravlenie*, is translated into English. *Upravlenie* is usually interpreted as "management" or "control." Hence, the problems of corporate governance in Russia are restricted within narrow spheres. I am afraid that this is the reason most economists regard corporate governance as factors related to only individual firms. However, corporate governance in Russia should be analyzed within corporate groups' governance, because these groups perform an essential role in the "real" Russian economy.

Corporate governance reflects phases of economic development, which necessitate government support and the institution of legislation. Hence, Russian corporate governance can be divided into three stages in the process of transformation from a socialist to a capitalist economy (Radygin, 2003, pp. 36-44).

During the first stage, prior to 1995, the concept of corporate governance was introduced in Russia. The Provisions on Joint Stock Companies (JSCs) were endorsed by the resolution of Republican Council of Ministers of Russia at the cabinet meeting on December 25, 1990. In the second stage, from 1995 and 2000, the first part of the Civic Code was enforced beginning on January 1, 1995, while the second part was established in 1995 and came into force from March 1996. In December 1995, the Law on JSCs was enacted. The Law on the Securities Market was also established in April 1996. A comprehensive program on the rights of depositors and shareholders was approved, based on the Decree of the President of the Russian Federation, enacted on March 21, 1996. According to the Criminal Code established in June 1996, an abuse of power such as insufficient supply of information at the point of issuing securities was legislated to be as criminal.

In the third stage, after 2000, both the Law on JSCs and the Law on the Securities Market were revised.[2] Other laws related to corporate governance were also enacted and revised. As shown in Table 1, under the regime of the Russian President Vladimir Putin, a number of laws were introduced, enacted or amended in order to establish and maintain corporate governance.

Except for the laws shown in Table 1, Law on the Protection of Rights and Legal Interests of Investors in the Securities Market was established in March 5, 1999. In addition, the Federal Committee of the Securities Market (now the Federal Service of Financial Markets) adopted a number of resolutions.[3] Specifically, the Committee determined many resolutions concerned with information disclosure.[4]

---

[2] Art. 42 of the Law on JSCs was revised on April 6, 2004, and enacted on July 1, 2004. Cl. 2, Art. 42, to the effect that the source of dividends is profits of the JSCs after payment of taxes, that is, net profits of firms, was added. The reason why net profits were exactly defined is that some firms tried to lower their net profits, which is the basis for dividends, regarding profits minus capital investments as net profits. For example, although an oil company, Surgutneftegaz, ought to pay dividends no less than 10 percent of net profits in its provisions, it was criticized for lowering the dividends of preference shares, because the definition of net profits was ambiguous. On July 7, 2004, another draft to revise the Law on JSCs was adopted in the first reading of the Lower House. In this draft, the following clause will be added: a holder of 90 percent of ordinary shares + 1 share of the JSCs or a holder of it with affiliated companies has the right to purchase the remaining shares, and the purchasing price should be the market prices approved by independent estimators (*Kommersant'*, September 24, 2004). But, in Russia there are many precedents whereby estimators estimate the value of shares unfairly.

[3] For instance, the standard on the occasion of establishing JSCs, and the standard of protocols of issuing additional shares and securities were endorsed by the Federal Committee of the Securities Market on September 17, 1996, and they were revised on November 11, 1998. The standard of protocols of issuing shares and securities on the occasion of reorganization of commercial organizations was approved on February 12, 1998, and was amended on November 11, 1998 (Medvedeva and Timofeev, 2003, p. 56).

[4] Although the Russian Law on JSCs prescribes the information presentation brought by JSCs (Art. 90), the information presentation to shareholders brought by JSCs (Art. 91), and obligatory disclosure of information brought by JSCs (Art. 92), the part of disclosure

Additionally, the Code of Corporate Conduct was endorsed by the Federal Government meeting under the guidance of the OECD. Then, it was approved by the Federal Committee of the Securities Market. The Code includes the principle of corporate conduct, the general meeting of shareholders, the board of directors, the executive organ of the firm, leaders of the firm, existing firm's behavior, disclosures concerning the firm, the control on financial and management activities of the firm, dividends, and the adjustment of confrontation among firms. The Code recommends the adoption of an appropriate form of management. For example, the recommendations endorsed the establishment of a strategic planning committee, a committee of auditors, a committee of staff and rewards, and a committee for the adjustment of confrontation among firms under the jurisdiction of a board of directors. The Code essentially is a trial for the introduction of an American-style of management.

Yet, has the Code of Corporate Conduct really been accepted by the Russian business community? According to data collected by the Russian Directors Institute and Managers Association during September and October 2002, only 15 percent of respondents were unaware of the Code of Corporate Conduct. However, only 19 percent responded that they had already established their own Code of Corporate Conduct (*Gotovnost' rossiiskikh kompanii*, 2002). Consequently, at least through autumn 2002, it was not reasonable to state that very many Russian firms actually practiced corporate governance, based on the Code of Corporate Conduct.

---

was intensified by revision and supplement of the Law on the Securities Market. Art. 30 of the Law prescribes the obligation to disclose the report of settlement of accounts each quarterly on the occasion of issuing securities and registering them. In addition, the Federal Committee of the Securities Market has determined a number of resolutions concerning the disclosure. For instance, the Committee endorsed the resolution about procedures and the sphere of the disclosure on the occasion of issuing shares on April 20, 1998 (OECD, 2002, p. 74). On August 12, 1998, regulations concerning the disclosure of important incidents and behaviors, which have an effect on the financial and economic activities of issuers, were endorsed by the Committee. On April 4, 2002, the Committee adopted the resolution to recommend the utilization of the Code of Corporate Conduct, in which detailed information concerning whether the firm obeys the provisions of the Code of Corporate Conduct is required to be disclosed in the annual report (Belikov, 2003, p. 2). The provision of disclosure concerning affiliated companies of open JSCs was endorsed by the resolution of the Committee on April 1, 2003. On July 2, 2003, the provision of disclosure concerning securities issued by issuers was also approved by the Committee.

An empirical study is important to understand the three stages of the development of corporate governance. The structure of ownership is one of the crucial factors, which has an effect on corporate governance. As shown in Table 2, as time progresses, the structure of ownership also has changed. From Table 2, it can be concluded as follows: first, at the beginning of privatization, insiders like employees and managers held more shares relative to outsiders; second, as time progresses, the holding share of insiders had declined superficially, which mainly resulted from the decrease in shares held by employees; third, the holding share of managers appear inclined to increase at least since 2001; and, fourth, the holding share of individuals, that is, the public at large, as outsiders, has tended to increase, while the holding share of banks remains at a relatively low level.

Here, we should remember that it is difficult to distinguish between the insiders and outsiders, since companies and individuals who have significant relationships with managers are included in the outsider group. Even if the result of the investigation shows that the structure of ownership has been transferred from the insiders to the outsiders superficially, this is not the "reality." Additionally, in Russia, it should be noted that the major owners are honorary owners who do not in actuality control their companies. Hence, it is necessary to specify the differing groups and their differing responsibilities with terms such as the "beneficial owners" or "ultimate owners."

To appreciate the de facto "reality" which influences corporate governance in Russia, attention to the structure of ownership is important. The problems with focusing attention on ownership should be argued at the level of the firm, and also at the industrial or state level. While the focus on ownership at the company level has the possibility of making corporate governance more straightforward and relatively compliant, this could encourage investment, and lead to economic growth. However, if emphasis is placed at the state level, it may effectively become an obstacle by, for example, disrupting competition, and distorting economic policies to the advantage of some parts of monopolistic and oligopolistic firms (World Bank, 2004, p. 90).

Table 3 shows the result of questionnaires concerning concentration of ownership. Of the 213 firms surveyed, the average holding share of the largest shareholders increased from 26.2 percent in 1995 to 27.6 percent in 1998. The average holding share of the 1st–3rd largest shareholders also increased from 40.4 percent to 44.5 percent. According to other data, these numbers seem to indicate that the rate of firms where one shareholder holds at least blocking shares also increased from 15 percent, when privatization began, to 33 percent in 2000 (Radygin, 2002, p. 108).

In this way, at the level of firms, the phenomenon of shares being concentrated in the hands of specific owners can be observed. It seems that a number of firms try to accrue as many shares as possible and thereby prevent interference from outsiders, since the legal rights of shareholders are relatively simple to subvert owing to court ineffectiveness. It is believed that the concentration of ownership stems from the limitations inherent in the enforcement of standards and laws enacted to protect the investor (La Porta, et al., 1998). On the other hand, concentration of ownership at the level of firms decreases the liquidity of shares, and makes it difficult for investors to monitor firms through the stock market.

Concentration of sales also has been increased. According to data collected from 1,700 firms during June and September 2003 by the World Bank, which covered 66 percent of sales and 22 percent of employment in industry, only 22 corporate groups had sales of more than 20 billion rubles or employ more than 20,000. These groups are shown in Table 4. Gazprom, Rosneft', UES, and Tatneft' are not included in it, because they are regarded as enterprises owned by the Federal or Regional Government. The sales of these 22 corporate groups accounted for 38.8 percent of the total. This sample amounts to 65.7 percent of total sales in industry, therefore, only 22 groups produce 25.5 percent of the total sales in industry. This indicates that concentration of sales has developed. On the other hand, concentration of employment is lower than the concentration of sales; however, 22 groups employ approximately 7 percent of the total employees in industry. Therefore, the role of corporate groups in the Russian economy is very important. This is the "reality."

CORPORATE GROUPS' GOVERNANCE

Some of Russia's excellent economists have already suggested that corporate groups are crucial factors and must be incorporated in any analysis of the Russian economy. Rozinskii (2002, p. 173) observes that a company as an economic entity has been disappearing. Kuznetsov, et al.

(2002, p. 71) indicate that official reports of the business accounts of firms do not reflect their actual conditions. Avdasheva (2004, p. 130) states that the characteristics of corporate groups concerned with corporate governance are as follows: (1) the relationship among shareholders in Russian corporate groups are opaque, and there are many cases in which a number of firms related to a parent company hold the controlling shares of subsidiary firms; (2) in addition, corporate groups try to change shareholders of their subsidiary firms in order to make the structure of ownership more opaque;[5] (3) in the holding companies, which are the fundamental type of Russian corporate groups, upstream firms hold shares of downstream firms to hide the controlling structure; (4) concentration of shares on shareholders or their affiliated companies has been developed; (5) the composition of the board of directors of firms belonging to corporate groups is characterized by their total control of them, not reflecting who and how many shares are held nominally; and, (6) in reality, the rate of ownership by insiders in share capital amounts is very high, that is, there are very few cases in which managers hold shares of their own firms openly. They are inclined to hold shares indirectly through their firms' network.

According to Avdasheva (2004, p. 123), corporate groups (business groups) include (1) firms producing products, (2) firm purchasing raw materials, (3) firms organizing the settlements between suppliers and consumers, and (4) banks financing firms' activities. She argues that factors to stimulate grouping or integration of firms consist of (1) imperfect protection against ownership, (2) exclusion of dual margins by the integration, and (3) reinforcement of the market control by horizontal integration (ibid., p. 126). However, in reality, transfer pricing and processing on commission played a greater role in the increase of vertical integration (ibid., pp. 127-129).

---

[5] After a parent company concentrates its shares on subsidiary firms and affiliated companies, and they come to be controlled by their parent company, the parent company begins to increase investment in subsidiary firms and affiliated companies in order to improve their performance. This is the same way in other countries (Shleifer and Treisman, 2003, p. 20).

TRANSFER PRICING

Transfer pricing enables a parent company to concentrate profits, to funnel profits to foreign subsidiaries, to transfer profits to subsidiaries located in well-established tax havens, or to simply evade taxes. The problems of transfer pricing are not limited to Russia. Expansion of foreign trade leads to the exploitation of transfer pricing to evade taxes. OECD has implemented a guideline for transfer pricing in 1979, and revised it into "Transfer Pricing Guidelines for Multinational Enterprises and Tax Administration" in 1995. Seen in this light, transfer pricing is a global concern. Specifically, in the case of Russia, not only in foreign trade, but also in transactions through domestic offshore companies and in the operations within corporate groups, transfer pricing is very widely employed, which ultimately results in a decrease in tax payments. It is said that against utilization of transfer pricing the tax authorities had made claims to banks, tobacco companies, Gazprom, Rosneft', TNK, and metallurgy holding companies in recent years (*Vedomosti*, p. A1, February 11, 2005).

Here, the case of transfer pricing in the oil and gas industry will be examined and discussed. First, utilizing transfer pricing, the question of how oil companies save their taxes will be explored. One of the well-known methods for minimizing the profit tax is the scheme of transfer pricing between domestic offshore companies and their parent companies or between affiliated companies registered in other preferential tax zones and their parent companies. This method could cut 16 percent points of 24 percent of the profit tax at the period of 2003. This scheme was realized as followed: (1) large-scale oil companies sold oil obtained from their subsidiary oil producing companies to their affiliated companies or dummy companies registered in domestic offshore or preferential tax zones by the price of one-half or one-third lower than the international price; (2) the affiliated companies sold the oil to their subsidiary refineries by the price of two thirds of the international prices, therefore, most profits were accumulated in the affiliated companies registered in domestic offshore or preferential tax zones.

The boom in establishing offshore companies by Russian residents and native firms occurred during 1992 and 1995. By the end of the 1990s, the number of offshore companies amounted to tens of thousands (ibid., p. 8). It can be confidently suggested that many major Russian firms are connected with offshore companies in, for example, Cyprus, Gibraltar, the British Virgin Islands, Luxembourg, and the Netherlands. Nevertheless, it is important to note that Russian firms do not necessarily need to establish offshore offices, due to the existence of offshore zones, or Free Economic Zones within the borders of Russia proper (see Table 5). In addition to Table 5, so-called "domestic offshore" zones, which provide exemptions for a portion of the fees due for profit and property taxes, were introduced in the Mordovian Republic and Chkotka Autonomous Okrug (AO). To the firms registered in Baikonul preferential tax treatment was also supplied, based on the agreement between the Russian and Kazakhstan Governments.

As a result, in 2002, the ratio of profit tax in the profits before taxation of the profit tax in case of Lukoil, excluding foreign taxation, amounted to 31.8 percent. Yukos's ratio was 12.9 percent, TNK's ratio was 14.3 percent, and Tatneft''s ratio was 24.2 percent. Sibneft''s ratio was 12.3 percent, and Surgutneftegas's ratio was 24.9 percent. The rate of profit tax was 24 percent in 2002. The reason why the above ratio of Lukoil exceeded 24 percent was that Lukoil had to pay additional tax to the tax avoidance, utilizing Baikonul preferential tax treatment.

According to Ivchikov (2002), the mechanism of transfer pricing can be shown as follows: (1) the head oil complex buys oil from its subsidiary oil mining companies based on the transfer price; (2) within 30 to 40 percent of the oil, one portion is exported, and the remaining oil is processed on the condition of paying commission; (3) the processed products of oil are sold to domestic and foreign markets. It is important to know that in the oil sector, processing on commission became widespread, simultaneously utilizing transfer prices. Avdasheva and Dement'ev (2000, p. 21) indicate that such a well integrated oil company like Lukoil began to utilize processing of commission more aggressively than an oil company like Sidanko, where a core company played a smaller role on controlling. In addition, it has been proposed that the ratio of processing of commission in all products in oil refineries amounted to 90 percent in 1999 (*Neftegazovaia vertikal'*, p. 66, No. 12, 2000).[6]

In the oil sector, the commissioner supplies oil to the oil factory in order to process it, and then receives some parts of processed products for

---

[6] As for the data concerning processing on commission, see Avdasheva, 2001, p. 101.

resale, leaving the other parts in the oil factory as a commission fee. In this process, the price of oil supplied to the oil factory is set as a transfer price. Generally speaking, transfer pricing in the oil sector is utilized between mining companies and refineries. Both belong to the same group. This transaction is realized through mediators or the parent company. In the case of mediators, mining companies sell oil to mediators at a lower price than the market price, and the mediators sell the oil to refineries belonging to the same group. If mediators are located in the domestic or foreign offshore or preferential tax zones, they can avoid profit tax. Some parts of processed products are retained on the condition of commission, while other parts are received by the mediators and resold.

It is well known that Yukos utilized such schemes. The transactions of Yukos were realized through organizations founded in Mordoviia, Kalmykiia, Evenkiia, where preferential taxes were applied, and Cheliabinsk, Sverdlovsk, Nizhegorod Oblast, where closed cities with preferential privileges were located (*Vedomosti*, p. A1, January 15, 2004). Yukos utilized transfer prices to reduce its own sales and increase profits of those organizations located in the preferential tax zones. However, Article 40 of the Tax Code, enacted on January 1, 1999, prescribes some cases in which the tax authorities have the right to control prices used in transactions. According to Clause 1, Article 40, the prices for taxes are applicable to the prices of goods, works, and services, specified by transaction parties, supposing these prices correspond to the level of market prices. In order to monitor whether this premise is fulfilled, the tax authorities are given the right to check the legitimacy of transaction prices only in the following circumstances: (1) transactions between "interdependent" persons; (2) good exchange (barter) transactions; (3) foreign trade transactions; and, (4) transactions where the level of prices used by the taxpayer for identical goods fluctuates by more than 20 percent in either direction over a relatively short period of time. This regulation was introduced to deter evasion of taxes through transfer pricing. However, procedures to realize this regulation are opaque, so this regulation has not proven especially effective. Additionally, the definition of market prices is also ambiguous; therefore, the application of this regulation is problematic.

In case of Yukos, since summer 2003, several executives were arrested, which aroused suspicions concerning management, therefore, a tax inspection was carried out. The number of companies concerned with oil transactions in the Yukos group amounted to 22, and Yukos itself. Closed

JSC Yukos-M is included in 22 companies. Although Yukos held only 4 percent of its shares on December 22, 2000, the tax authorities regarded it as an affiliated company of Yukos and inspected it, because all sales of Yukos-M were brought about by the transactions with Yukos. Even if there are no relations with more than 20 percent of capital, some companies are considered as affiliates. For instance, in 2000, Yukos sold oil to Yukos-M at the price of 750 rubles per ton, then Yukos consigned Yukos-M to export the oil at the price of more than 4,000 rubles (*Vedomosti*, p. A1, July 23, 2004). As a result, the tax authorities calculated that the sum of the unpaid taxes of Yukos in 2000 amounted to 98 billion rubles, of which 47 billion rubles were unpaid taxes and the remaining was a surcharge and a penalty.

Concerning the case when a parent company *per se* plays a role in setting transfer prices, the instance of oil company Rosneft' is well known. In January 2001, the Moscow Arbitration Court acknowledged that the contract, in which a subsidiary of Rosneft', Purneftegaz sold 9.3 million tons of oil to Rosneft' at the price of 1,110 rubles per ton, was invalid (*Vedomosti*, p. B2, March 14, 2002). This suit was filed by shareholders of Purneftegaz, because this price was too low in comparison with market prices. The Court considered that the market price at that time was approximately 2,000 rubles per ton, therefore, this condition of the contract was disadvantageous to Purneftegaz.

Generally speaking, a domestic refinery purchases oil from an organization from the same group, an independent oil producer under contract, or a free market (*Neftegazovaia vertikal'*, p. 65, No. 12, 2000). According to the data investigated by this source, 24 refineries in 1999, the average price of oil within the internal transaction, that is, the average transfer price amounted to 600 to 800 rubles per ton, while the oil price was based on the independent oil producer, as 1,350 to 1,600 rubles per ton, and the oil price at the free market was 3,150 to 3,200 rubles per ton (ibid., p. 66). In case of Yukos, the free sale price of its subsidiary mining companies exceeded five times the transfer price of their affiliated refineries. Lukoil-West Siberia and KomiTEK supplied oil to unaffiliated refineries based on the direct contract at two times a higher price than their transfer prices. Although Surgutneftegaz set its transfer price three times higher than other oil companies' transfer prices, it added 17 percent of the transfer price to the price based on the direct contract, and 25 percent to the price for a free market.[7] At this point, it should be noticed that the purchasing price of oil at refineries collected by the Statistical Service is based on a market price, however, the volume amounts to several percent of the real oil volume received by refineries. Concretely, the price was almost 3,300 rubles per ton both in 2000 and 2001 (Volkonskii and Kuzovkin, 2002, p. 23).

The reason why transfer pricing was widely utilized is that the sales of mining minerals were targets of taxation until the introduction of the mining tax on mineral resources, integrating using fee on underground resources, deduction for reproducing minerals and resource bases, and excise taxes on oil and gas in the beginning of 2002. In case of using fee on underground resources, the target of taxation was the sum of sales of mining minerals. In calculating deduction for reproducing minerals and resource bases, the target of taxation was the sales of minerals and resources received by mining companies. On the other hand, the mining tax on mineral resources is determined by the volume of mining oil. The tax rate is fixed against the volume of mining, therefore, the incentive to utilize transfer pricing to underestimate the sales has declined.

As for a profit tax, if a company can pretend to show smaller sales in its accounts, it results in reducing the profit tax. Nevertheless, it can be concluded that the incentive to underestimate sales has also faded, because in 2002 the tax rate of the profit tax was reduced from 35 percent (11 percent for the federal budget, 24 percent for the regional budgets) to 24 percent (7.5 percent, 16.5 percent respectively). In addition, since 2004, regional preferential privileges concerned with the profit tax have been restricted within 4 percent points of 24 percent. Of course, after the Yukos scandal, oil companies are exposed in public, therefore, it became difficult for them to utilize transfer pricing to save taxes.[8]

---

[7] We should not that the ratio of materials supplied from the same group's mining companies in the total processed oil of refineries registers a difference of 45 percent (ONAKO) versus 100 percent (Yukos). Comparing the ratio of refineries supplied from their same groups' mining companies in 1996, these ratios of SIDANKO, Rosneft', and ONAKO amounted to only 20 to 30 percent (Kriukov, 1998, p. 201).

[8] Nevertheless, in other sectors, transfer pricing is utilized frequently. According to the report "about the estimation of the level of payment of taxes concerning a large scale of metallurgical companies," they intended to reduce taxes, utilizing processing on commission, transfer pricing, regional subsidies, and mediators registered in the preferential tax zones. For instance, it is indicated in this report that "North Steel" reduced taxes by 1.2 billion rubles, purchasing materials from the mediator registered in Kalmykiia. On the

As for the gas industry, transfer pricing was frequently utilized. *Vedomosti* stated that "Even Vlagimir Putin knew that the Gazprom exported gas at a price one-half to one-third the market price," then, President Putin asked the president of the Gazprom where the price difference had gone, in November 2001 (*Vedomosti*, p. A1, March 13, 2002). According to *Ekspert*, the Administration of Iamaro-Nenets AO received gas as a fee for using underground resources from the Gazprom, at the price of 2 to 3 dollars per 1,000 cubic meters and sold it to ITERA, holding deep connection with the Gazprom, at the same price (*Ekspert*, p. 14, No. 21, 2001). Although the Gazprom was damaged by one billion dollars in a year in this scheme, most of the loss flowed out through the ITERA, and some persons must have made a profit.[9]

Although the Gazprom seemed to utilize transfer pricing not only in the transactions of export, but also in the domestic operation, it is difficult to collect data concerning domestic transfer pricing.

To reinforce the regulations on transfer pricing, the Russian Federation Ministry of Finance has discussed the draft of the Law on the Control on Transferring Prices.[10] This draft was passed in the first reading of the Lower House in 2000, but failed to become law. Currently, the Ministry is still attempting to revise the draft for re-submission to the Diet. At the end of 2004, the Ministry proposed to the Government the draft of the Law, in which the rights to assume control over all the transactions among affiliated companies would be provided to the Ministry of Finance. Another method to deter transfer pricing is to introduce the tax payment system based on consolidated accounts (*Vedomosti*, March 25, 2004). In either case, because of transfer pricing, profits are redistributed among corporate groups, and therefore, it seems very difficult to estimate the real profits from the profits of individual firms.

From the point of "corporate groups' governance," it can be concluded that corporate groups strengthen their position through holding or even hoarding shares internally, i.e., the insiders amass as many shares as

---

other hand, the utilization of transfer pricing to avoid an excise tax has been at issue in these years, for example, in the tobacco sector. The excise tax on oil *per se* was abolished in 2002.

[9] In this scheme, the payment in goods to the Administration was forbidden in 2001.

[10] In Kazakhstan, the Law on the State Control on the Application of Transfer Prices was enacted in 2001. In 2003, the joint instructions of the Tax Committee under the Ministry of Finance and the Customs Control Service were published.

possible which effectively diminishes the potential of outsiders to have any influence on corporate strategy. However, it is meaningless to distinguish insiders from outsiders, at least on a superficial level. Individuals and firms regarded as outsiders include stakeholders who could have the support of the insiders, like managers. Therefore, it is necessary to distinguish between what might be considered superficial "insiders' ownership" and what in actuality is genuine "insiders' control." Without making unambiguous distinctions between "ultimate owners" and "beneficial owners," arguments about the challenges and problems associated with "corporate groups' governance" are very much weakened.

As a result of considering corporate governance as only a problem concerning each firm, injustice related to corporate groups is overlooked. Minority shareholders are pressured in various ways, and the profits of individuals controlling corporate groups and the profits of corporate groups as a whole are increasing. This is a history of Russian firms in the process of conversion from a socialist economy to a capitalist one. At the beginning of grouping firms, the dilution of shares could avoid being controlled by a third party, and their integration was stimulated by the mutual holding of shares and by holding the shares of parent companies by affiliates (The Minority Squeeze, 2004, p. 40). In addition, "reverse stock split" could prevent investors from purchasing shares by increasing the price of "one share" by the integration of several shares. Utilizing transfer pricing, corporate groups could redistribute profits among them, evade taxes, and manipulate share prices of specific firms. This is the most crucial issue in relation to corporate governance.

From the point of "corporate groups' governance," the Law on Holding Companies, the Law on Affiliated Companies, and the Law on Transferring Prices are very important in establishing "corporate groups' governance." However, even now, neither of the laws has been enacted.

As for the draft of the Law on Holding Companies, President Putin vetoed it in 2000.[11] In 2001, another draft was proposed but later rejected by the Upper House. Although the draft was re-submitted to the Lower House at a later time, the draft has not finished its run through committee procedures as of October 2004. Therefore, at this time, holding companies are regulated by the appendix of the Presidential Decree of November 16,

---

[11] The draft of the Law on Holding Companies presented to the third reading of the Lower House in 1999 can be seen in Gorbunov, 2003, pp. 197-206.

1992, called the "Temporary Provisions concerning Holding Companies Transferred from State-Owned Enterprises to Joint Stock Companies."

The Law on Affiliated Companies has also not been established. According to Articles 105 and 106 of the Civic Code, if a firm holds more than half of the shares of another firm, the latter firm is called a subsidiary firm, and if it holds more than 20 percent of shares, it is called a "dependent" firm (*zavisimoe obshchestvo*). The *zavisimoe obshchestvo* can be also translated as an "affiliated" firm. Therefore, dependent or affiliated companies include not only subsidiary firms, of which more than 50 percent of shares are held by their parent firms, but also other firms, of which 20 to 50 percent of shares are held by their parent companies. Besides, firms controlled by the same managers can be seen as affiliated companies (*Kommersant'*, October 1, 2004). According to Article 20 of the Tax Code, "dependent" is defined as more than 20 percent commitment of an organization to other organizations directly or indirectly. Not only a direct relationship, but also indirect ones are recognized as "dependent." However, the relationship between the existence of a "dependent" relationship and the responsibility to pay taxes is ambiguous. Therefore, it is necessary to define affiliated companies clearly and reinforce regulations on corporate groups.[12]

---

[12] There are other factors to consider about "corporate groups' governance." Radygin argues that one of the dominating motives of a "beneficial owner" in concealing information of the real owners of some share is protection of assets acquired recently and far from always completely legally (Radygin, 2003, p. 5). Russian companies frequently use trusts for purposes of minimizing their tax exposure. For instance, a "beneficial owner" located in a country with a high taxation regime transfers title of his property to a trustee resident in an offshore zone. In this case, the trustee would pay property tax at lower rates, and the "beneficial owner," in turn, would pay taxes only on profits gained as a result of using his property. Apparently, the reason offshore companies were established was to conceal profits and capital, evade taxes, and protect assets. All things considered, it is true that offshore companies can hide "beneficial owners." Hence, from the point of corporate governance, the regulations on the relationship between offshore companies and Russian residents or native firms should be strengthened. One of the regulations is introducing the legal requirement of beneficial ownership information disclosure by offshore companies (ibid., p. 5). Offshore companies are concerned with trust. Trust schemes appear to follow a prescribed path: property to be transferred in trust is first passed over to an offshore company, after which the shares of that company are placed in trust (ibid., p. 11). The concept of trust appeared in the Law on Banks and Banking Operations in Russian Republic of the Soviet Union, established in December 2, 1990. In Article 5 of the Law, "trust management" (*doveritel'noe upravlenie*) means the operation of accepting money and controlling

To disclose "beneficial owners," that is, to establish "corporate groups' governance," the regulations on information disclosure are very important. According to Article 30 of the Law on the Securities Market, an issuer of securities must disclose the list of holders who have more than 20 percent of its share capital and juristic persons which have been issued more than 20 percent. Every holder who has more than 20 percent of any class of securities also must disclose the information about their securities. If more than 5 percent of shares over the crucial 20 percent are increased or decreased, the holder is also obligated to disclose the information of his or her securities. The information of a newcomer who holds more than 25 percent of the shares should also be disclosed. In the protocol of securities, the information about persons holding no less than 5 percent of share capital and the list of all juristic persons holding more than 5 percent of share capital must be disclosed (Article 22). There are other regulations on information disclosure such as the resolutions of the Federal Committee of the Securities Market, the Law on Violation against Administrative Laws (Article 15.19), and the Law on Regulations against Monopolistic Activities and Competition in Commodity Market (Article 18), established on March 22, 1991. However, currently, information disclosure has not been sufficient enough to expose the "beneficial owners."

CONCLUSIONS

The policy to disclose "beneficial owners" and "ultimate owners" is the foundation from which to establish "corporate groups' governance." Radygin proposes the following concrete measures: (1) clarification of the

---

securities on the basis of the trust of a client. In the Decree of the President on July 1, 1992, the expression of "trust" had appeared in Clause 6. On October 24, 1993, the Presidential Decree on Beneficial Ownership (trust) was published. At the beginning of 1994, the Russian Federal Fund of Property concluded the agreements with trust contracts, which amounted to approximately 80. The procedures and conditions of trust management of property were established by Article 209 of the Civic Code, enacted from January 1, 1995. The order of trust management of credit organizations and banks is regulated by the Law on Banks and Banking Operations and the Instruction on the Order of Exercising Trust Management Transactions and Accounting for Such Transactions by Credit Organizations in the Russian Federation (ibid., p. 19). However, even now the absence of the concept of trust in Russian law is crucial for purposes of identifying "beneficial owners" (ibid., p. 19). This is one of the reasons why "corporate groups' governance" has been fully developed in Russia.

legal concept of "beneficial owners;" (2) the strict implementation of laws and regulations concerned with trusts; (3) making assets and the control of shares transparent; (4) faithfully adhering to the Law on Firms and the Anti-Monopolist Law; (5) restricting offshore transactions through regulations on banks; and, (6) introducing International Accounting Standards and Generally Accepted Accounting Principles (Radygin, 2003, pp. 24-28).

To establish "corporate groups' governance," not only policies of information disclosure, but also the establishment of the following laws is crucial: (1) the Law on Holding Companies; (2) the Law on Affiliated Companies; and, (3) the Law on Transfer Pricing. Lastly, the "reality" of the Russian economic climate should be scrutinized, and then measures for managing it accordingly should be taken.

REFERENCES

Afanas'ev, M., P. Kuznetsov, and A. Fominykh, "Korporativnoe upravlenie glazait direktorata," *Voprosy ekonomiki,* 5:84-101, 1997.

Aukutsionek, S., and R. Kapeliushnikov, "Ownership Structure of Russian Industrial Enterprises in 2001," *The Russian Economic Barometer,* **10,** 3:10-17, 2001.

Avdasheva, S., "Integratsionnye protsessy v promyshlennost': institutsional'noe razvitie," in T. Dolgopiatova, ed., *Rossiiskaia promyshlennost': institutsional'noe razvitie, vyp. 1.* Moscow, Russia: Vysshaia shkola ekonomiki, 2002, 91-111.

Avdasheva, S., "Biznes-gruppy v rossiiskoi promyshlennosti," *Voprosy ekonomiki,* 5:121-134, 2004.

Avdasheva, S., "Daval'chestvo v rossiiskoi promyshlennosti: prichiny i rezul'taty," *Voprosy ekonomiki,* 6:100-113, 2001.

Avdasheva, S., and V. Dement'ev, "Aktsionernye i neimushchestvennye mekhanizmy integratsii v rossiiskikh biznes-gruppakh," *Rossiiskii ekonomicheskii zhurnal,* **1:**13-27, 2000.

Belikov, I., "Adoption of Russian Code of Corporate Conduct: Accomplishments and Problems," paper presented at Russian

Corporate Governance Roundtable Workshop "Implementation and Enforcement of Disclosure Rules," OECD, October 2–3, 2003 [http://www.oecd.org/].

Berglöf, E., and E.-L. von Thadden, "The Changing Corporate Governance Paradigm: Implications for Transition and Developing Countries," *World Bank Working Paper*, **263**, 1999.

Bevan, Alan A., Saul Estrin, Boris Kuznetsov, Mark E. Schaffer, **Manuela Angelucci, Julian Fennema, and Giovanni Mangiarotti,** "The Determinants of Privatized Enterprise Performance in Russia," *William Davidson Working Papers*, **452**, 2001 [http://www.wdi .umich.edu/].

Brock, G., "Public Finance in the ZATO Archipelago," *Europe-Asia Studies*, **50**, 6:1065-1081, 1998.

Brock, G., "The ZATO Archipelago Revisited: Is the Federal Government Loosening Its Grip? A Research Note," *Europe-Asia Studies*, **52**, 7:1349-1360, 2000.

Corporate Governance in Russia: Investor Perceptions in the West and Business Reality on the Ground. London/Moscow: SRU Limited, Expert Information Group, 2003. [http://www.gateway2russia.com/]

Dolgopiatova, T., "Modeli i mekhanizmy korporativnogo kontrolia v rossiiskoi promyshlennosti *(opyt empiricheskogo issledovaniia),*" *Voprosy ekonomiki*, 5:46-60, 2001

Dolgopiatova, T., "Corporate Control in the Russian Companies: Models and Mechanisms," Preprint Seriia WP 1 2002/05, Moscow, Russia: Higher School of Economics, 2002 [http://www.hse.ru/science/ preprint/WP1_2002_05.htm].

Dolgopiatova, T., "Stanovlenie korporativnogo sektora i evoliutsiia aktsionernoi sobstvennosti," Preprint Seriia WP 1, Moscow, Rossia: Vysshaia shkola ekonomiki, 2003 [http://new.hse.ru/C3/C18/ preprintsID/default.aspx?filter=WP1].

Earle, John S., and Saul Estrin, "Privatization and the Structure of Enterprise Ownership," in Brigitte Granville, Peter Oppenheimer, eds.,

Russia's Post-Communist Economy. Oxford, UK: Oxford University Press, 2001, 173-212.

Gorbunov, A., *Dochernie, kompanii, filialy, kholdingi.* Moscow, Russia: GLOBUS, 2003.

Gotovnost' rossiiskikh kompanii k prakticheskomu bnedreniiu rekomendatsii Kodeksa korporativnogo povedeniia. Moscow, Russia: Rossiiskii institut direktorov, Assotsiatsiia menedzherov, 2002 [http://www.rid.ru/research.php?id=242].

Ivchikov, N., "Transfertnye tseny v rynochnoi ekonomike," *Finansovyi menedzhment,* 3, 2002.

Iwasaki, I., "The Governance Mechanism of Russian Firms: Its Self-enforcing Nature and Limitations," *Post-Communist Economies,* **15**, 4:503-531, 2003.

Judge, W., and I. Naoumova, "Corporate Governance in Russia: What Model Will Follow?," *Corporate Governance,* **12**, 3:302-313, 2004.

Kapeliushnikov, R., "The Largest and Dominant Shareholders in the Russian Industry: Evidence of the Russian Economic Barometer Monitoring," *The Russian Economic Barometer,* **9**, 1:9-46, 2000.

Kostikov, I., "Improving Corporate Governance in Russia," presentation at ICGN 2003 Annual Conference Amsterdam, International Corporate Governance Network, July 9–11, 2003 [http://www.icgn .org/conferences/2003/index.php].

Kriukov, V., *Institutsional'naia struktura neftegazovogo sektora: problemy i napravleniia transformatsii.* Novosibirsk, Russia: IEiOPP SO RAN, 1998.

Kuznetsov, P., G. Gorobets, and A. Fominykh, "Neplatezhi i barter kak otrazhenie novoi formy organizatsii promyshlennosti v Rossii," in *Predpriiatiia Rossii: korporativnoe upravlenie i rynochnye sdelki.* Moscow, Russia: Vysshaia shkola ekonomiki, 2002, 28-79.

Kuznetsova, O., and A. Kuznetsov, "The State as a Shareholder: Responsibilities and Objectives," *Europe-Asia Studies,* **51**, 3:433-445, 1999.

La Porta, R., F. Lopez-de-Silanes, and A. Shleifer, "Corporate Ownership Around the World," *NBER Working Paper Series,* **6625,** 1998 [http://www.nber.org/papers/w6625.pdf].

Medvedva, T., and A. Timofeev, "Iuridicheskie aspekty issledovaniia sprosa na instituty korporativnogo upravleniia," in *Razvitie sprosa na pravovoe regulirovanie korporativnogo upravleniia v chastnom sektore.* Moscow, Russia: Moskovskii obshchestvennyi nauchnyi fond, 2003, 83-111 [http://www.ecsocman.edu.ru/db/msg/214039.html].

Nasheri, Hedieh, *Economic Espionage and Industrial Spying.* London, UK: Cambridge University Press, 2005.

OECD, *White Paper on Corporate Governance in Russia,* 2002 [http://www.oecd.org/].

Radygin, A., "Post-Privatization' Corporate Governance in Russia: Singular Path or Typical Transition Trajectory?," paper presented at International Conference of Groupe Transition Developpement "Eastern Transition Trajectories: Measurement, typologies, differentiation, interpretations," GTD-Grenoble, December 10–11, 1999.

Radygin, A., "Korporativnoe upravlenie v Rossii: ogranicheniia i perspektivy," *Voprosy ekonomiki,* 1:101-125, 2002.

Radygin, A., "Beneficial Ownership information Disclosure," paper presented at Russian Corporate Governance Roundtable Workshop "Implementation and Enforcement of Disclosure Rules," OECD, October 2–3, 2003 [https://www.oecd.org/].

Radygin, A., and S. Arkhipov, "Sobstvennost' i finansy predpriiatii v Rossii (empiricheskii analiz)," *Obzor ekonomiki Rossii,* 2:40-56, 2001 [http://www.recep.ru/phase4/pdf's/ret2001q2r.pdf].

Radygin, A. D., R. M. Entov, A. E. Gontmakher, I. V. Mezheraups, and M. Iu. Turuntseva, *Ekonomiko-pravovye faktory i ogranicheniia v stanovlenii modelei korporativnogo upravleniia.* Moscow, Russia: IEPP, 2004.

Radygin, A., B. Gutnik, and G. Mal'ginov, "Postprivatizatsionnaia struktura aktsionernogo kapitala i korporativnyi kontrol': 'kontrrevoliutsiia upravliaiushchikh'?," *Voprosy ekonomiki,* 10:47-70, 1995.

Roberts, G., "Convergent Capitalism? The Internationalisation of Financial Markets and the 2002 Russian Corporate Governance Code," *Europe-Asia Studies,* **56,** 8:1235-1248, 2004.

Rozinskii, I., "Mekhanizm polucheniia dokhodov i korporativnoe upravlenie v rossiiskoi ekonomike," in *Predpriiatiia Rossii: korporativnoe upravlenie i rynochnye sdelki.* Moscow, Russia: Vysshaia shkola ekonomiki, 2002, 25-35.

Shiobara, T., *The Reality of Russia's Economy.* Tokyo, Japan: Keio University Press, forthcoming, 2006.

Shleifer, A., and D. Treisman, "A Normal Country," *NBER Working Paper Series,* **10057,** 2003 [http://www.nber.org/papers/w10057].

Sprenger, C., "Ownership and Corporate Governance in Russian Industry: A Survey," *EBRD Woking Papers,* **70,** 2001.

Stravoitov, M., "Aktsionernaia sobstvennost' i korporativnye otnosheniia," *Voprosy ekonomiki,* 5:61-72, 2001.

The Minority Squeeze Out Law: Problems with the Proposed Amendments, Hermitage Capital Management, 2004.

Ushakov, D., *Ofshornye zony v praktike rossiiskikh nalogoplatel'shchikov.* Moscow, Russia: Iurist', 2002.

Vsemirnyi bank, "Sobstvennost' i kontrol' predpriiatii," *Voprosy ekonomiki,* 8:4-35, 2004.

Volkonskii, V., and A. Kuzovkin, "Tseny na toplivo i energiyu. Investitsii. Biudzhet," *Ekonomika i matematicheskie metody,* **37,** 2:22-37, 2001.

Volkonskii, V., and A. Kuzovkin, "Neftianoi kompleks: finansovye potoki i tsenoobrazovanie," *Ekonomist,* 5:21-32, 2002.

World Bank, *From Transition to Development: A Country Economic Memorandum for the Russian Federation,* Draft, 2004 [www .worldbank.org.ru].

Table 1. Main Laws Related to Corporate Governance and Protection of Ownership, Which Were Revised or Enacted during 2000 and 2003

Laws	Contents	Plus	Minus
Revision and Supplements of the Law on JSCs (August 7, 2001, March 21, 2002, October 31, 2002) and Revision of the Law on Protection of Rights and Legal Interests of Investors in Securities Markets (December 9, 2002)	Prohibition against issuing shares for a part of shareholders. The compulsory integration of shares as the method of throwing minorities out is restricted. The proportional exercising of the rights of shareholders on the occasion of reorganization (separation) is introduced. The procedures for controlling organizations are specified.	Most methods which allowed for discrimination against the rights of shareholders were prohibited.	Revision was delayed for several years. Important issues such as clarification of rules of merger and regulations of division of shares are unresolved.
Revision and Supplements of the Law on the Securities Market (December 28, 2002)	New requests for information content to be disclosed by the issuers. Simplification of regulation procedures for issuing securities of the closed JSCs. Introduction of responsibility for price manipulation. Regulations on option trading.	Partial improvement of opaque laws and regulations	Unfair expansion of full power of regulation authorities. Introduction of compulsory utilization of financial consultants.
New Law on Insolvency (Bankruptcy) (October 26, 2002)	Strengthening protection of the rights of creditors (first of all, the state). Expansion of the sphere of the rights of the owners in good faith, who own firms in debt in the procedures of bankruptcy. Alteration of the status of arbitration receivers and governmental institutions. Introduction of new proce-dures for sound finance.	Establishment of obstacles to hostile mergers, utilizing the mechanism of bankruptcy.	Retaining theft of ownership in an altered way and conditions to corruption.
Law on Investment Funds (November 29, 2001)	Legal basis for activities of investment funds, their controlling companies, and special depositors is prescribed.		It is necessary to develop rules of trust by closed and open investment funds and investment funds with holdings.

Law on Privatization of Public Assets (December 21, 2001)	Alteration of a set of methods enabled to abolish minority holdings and unemployed capital.	In the sphere of abolishment of stocks of uncontrolled ownership, the basis to realize the concept of the control of federal assets (2003) has been brought about.	It is necessary to show that new laws are effective in a few years and to adopt the law on public assets.
Law on Public Unitary enterprises (November 14, 2002)	The level of protection of national interests is improved drastically.	Stimulating refusal of the form of organizations and rights of unitary enterprises.	Ten years lag (this issue has been discussed since 1993. Notes concerning this law had already existed in the Civil Code in 1995).
Revision of Criminal Code (2001)	This revision lays criminal responsibility on leaders of JSCs. This responsibility involves the violation of protection of rights and interests of investors.	These regulations concerned with criminal responsibility have original significance in relation to rights accorded in Russia.	
New Code on Violation of Administrative Law (December 30, 2001)	This code includes the regulations on relative responsibility in the sphere of corporate governance in securities markets.		Absence of transparent measures in evaluating sanctions in violation of the law.
Code of Arbitration Procedures (enacted in September 1, 2002) and Code of Civic Procedures (enacted February 1, 2003)	Adoption of dual system in the trial of lawsuits concerning firms (Arbitration courts and general courts).		Collision with jurisdiction in the sphere of protection of the rights of shareholders is unresolved. Proper alteration to the Code of Arbitration Procedures is necessary.

Sources: Radygin, et al., 2004, pp. 278-280.

Table 2. Structure of Ownership in Russia, Based on Sample Investigation

	Afanas'ev, et al. [1]			Worldbank Sample [2]	Radygin, et al. [3]		Starovoitov [4]		Dolgopiatova [5]			Spenger [6]
	Jul.–Aug., 1994	Dec., 1995 –Jan., 1996	1996	1994	Apr., 1994	Jun., 1995	1995	1997	1995	1998	2000	1999
Insiders	**66**	**58**	**55**	**66.1**	**62**	**56**	**58.5**	**51.6**	**49.8**	**40.1**	**39.3**	**46.2**
Managers	22	15	12	19.6	9	13	10	12.1	7.8	9	9.5	14.7
Employees	44	43	43	46.2	53	43	48.5	39.5	42	31.1	29.8	31.5
Outsiders	-	-	**26**	**18.9**	**21**	**33**	**31.7**	**41.3**	**37.9**	**49.3**	**49.6**	**42.4**
Individuals	6	8	8	5.9	-	-	9.6	13.2	13.5	18.6	19	18.5
Non-financial firms	-	-	3	6.7	-	-	8.1	12.9	12	13.9	14.9	13.5
Banks	-	-	2	1	-	-	1.6	1.2	1.6	1.3	1.2	1
Investment funds	5	5	5	4.5	-	-	7.2	4	9.0ᵃ	11.8	11.2	3.9
Foreign investors	-	-	2	0.4	-	-	1.7	5.1	1.8	3.7	3.3	2
Holding companies/ Investment firms/Trust	-	-	6	-	-	-	3.5	4.9	-	-	-	3.5
State	**12**	**9**	**9**	**15**	**17**	**11**	**9.5**	**6.5**	**9.7**	**8.4**	**8.9**	**7.1**
Others	**11**	**20**	**10**	-	-	-	**0.3**	**0.6**	**2.6**	**2.2**	**2.2**	**4.3**
The number of samples	88	312	259	235	-	-	138	139	277	277	277	139

Table 2. (continued)

	Bevan, et al. [7]	Radygin & Arkhipov [8]	Aukustionek & Kapeliushnikov [9]				Radygin, et al. [10]		Dolgopiatova [11] d		
	2000	2000	1995	1997	1999	2001	1999	2002	2000	2001	2002
Insiders	**62.3**[b]	**27.6**	**54**	**52**	**50**[c]	**50**[c]	**37.8**	**32.7**	**52.5(-15.0)**	**48.2(-6.6)**	**31(-14)**
Managers	17.7	7.2	11	15	15	19	9.4	17	17.8(4.9)	21.0(9.8)	9(1)
Employees	34.5	20.4	43	37	34	28	28.4	15.7	34.7(-19.9)	27.2(-16.4)	22(-15)
Outsiders	**32**	**55.4**	**37**	**42**	**42**	**42**	**46.5**	**55.4**	**41.8(21.6)**	**43.9(7.8)**	**41(7)**
Individuals	-	15.2	11	15	20	22	22.2	27.4	19.3(9.6)	25.3(13.5)	14(2)
Non-financial firms	-	15.2	16	16	13	12	20.3	25.2	15.1(7.1)	11.3(-3.7)	22(9)
Banks	-	2.2	1	1	1	1	1.1	1.3	4.5	0.8	2(1)
Investment funds	-	4.4	4	4	3	3	-	-	-	6.5	1
Foreign investors	-	4.7	1	2	2	0	2.9	1.5	2.9(2.6)	0.4	2
Holding companies/ Investment firms/Trust	-	6.2	4	4	3	4	-	-	-	-	-
State	**5.7**	**12.8**	**9**	**7**	**7**	**7**	**11.4**	**8.2**	**5.7(-6.6)**	**7.9(-1.2)**	**28(7)**
Others	-	**4.2**	-	-	-	-	**4.3**	**3.7**	-	-	-
The number of samples	364	201	136	135	156	154	60	60	more than 350	more than 150	-

Notes: [a] Including investment firms. [b] The sum of managers and employees amounts to 52.2 percent. It seems that this figure includes other factors such as affiliated companies, but details are unknown. [c] Including holding shares of subsidiary firms (in 1999, 1 percent, in 2001, 3 percent). [d] Figures in parentheses express the changing rate (percent points). In 2000, the rate is related to the period during 5–8 years, in 2001, it is concerned with the time during 7 years, and in 2002, the rate expresses the result of 6 years. The figures for individuals include other factors. The figures for the bank in 2000 contain investment funds. In 2002, the samples are limited only to military firms.

Sources: [1] Afanas'ev, et al., 1997, p. 87. [2] Earle and Estrin, 2001, p. 183. [3] Radygin, et al., 1995, p. 53. [4] Starovoitov, 2001, p. 63. [5] Dolgopiatova, 2001, p. 47, Dolgopiatova, 2002, p. 6. [6] Sprenger, 2001, p. 4, the original is Kapeliushnikov, 2000. [7] Bevan, et al., 2001, p. 19. [8] Radygin and Arkhipov, 2002, p. 44. [9] Aukutsionek and Kapeliushnikov, 2001, p. 11. [10] Radygin, et al., 2004, p. 225. [11] Dolgopiatova, 2003, p. 47.

Table 3. Concentration of Ownership (survey of 213 firms)

(percent of total capitals)

	1995		1998		2000 (estimate)	
	Average	Mean	Average	Mean	Average	Mean
The largest shareholder	26.2	22	27.6	23	28.8	24.2
1st–3rd largest shareholders	40.4	40	44.5	44.4	46.5	46.3

Sources: Dolgopiatova, 2001, p. 47.

Table 4. Ranking of Employment and Sales among 22 Corporate Groups

Rank by Employment	Employment	Sales (1,000 rubles)	Managed by	Organization	Rank by sales
1	168,966	64,825,452	Deripaska	Base Element	11
2	168,554	202,629,008	Abramovich, Shvidler	Sibneft'/Millhouse	2
3	167,223	111,593,552	Kadannikov	Avtovaz	7
4	143,437	70,276,496	Popov, Mel'nichenko, Pumpianskii	MDM	10
5	136,868	474,973,216	Alekperov, Maganov, Kukura	Lukoil	1
6	121,901	78,224,152	Mordashov	Severstal'	9
7	111,692	137,194,080	Potanin, Prokhorov	Interros	5
8	101,091	52,412,024	Abramov	Evraz	13
9	94,047	121,121,744	Veksel'berg, Blavatnik, Balasskul	Renova/ Access Industries	6
10	93,271	149,226,576	Khodorkovskii, Lebedev	Yukos	4
11	74,933	33,221,580	Makhmudov, Kazitsin	UGMK	16
12	65,325	163,129,392	Bogdanov	Surgutneftegaz	3
13	56,892	57,199,712	Rashinikov	Magnitogorsk steel	12
14	53,932	30,854,502	Zuzin	Mechel	17
15	47,326	38,951,240	Lisin	Novolipetsk steel	15
16	41,698	20,439,996	Smushkin, Zingarevich	Ilim Palp	19
17	41,046	40,611,844	Takhaudinov	Tatneft'	14
18	38,490	106,713,016	Fridman, Khan	Alfa	8
19	35,935	15,113,239	Ivanishvili, Gindin	Metalloinvest	21
20	35,384	10,265,729	Bendukidze, Kazbekov	OMZ	22
21	20,272	26,946,746	Evtushenkov, Novitskii, Gotscharuk	Sistema	18
22	12,704	20,254,446	Iakobashvili, Plastinin, Dubinin	Vimm Bill' Dann	20
total	1,830,987	2,026,177,742			

Sources: Vsemirnyi bank, 2004, p. 12.

Table 5. Main Places Applied for Preferential Tax Treatment

1994	Established a Free Economic Zone 《Ingushetiia》, based on the resolution of the Government of the Russian Federation on June 19, 1994
1995	Established a Free Economic Zone 《Kalmykiia》[a]
1996	Began to supply preferential tax treatment to ZATO[b]
1997	Established a Free Economic Zone 《Uglich》
	Established a Free Economic Zone 《Altai》, based on the Law on the Free Economic Zone in Altai krai
	Introduced preferential treatment to Smolensk
	Abolished a Free Economic Zone 《Ingushetiia》
2001	Abolished preferential tax treatment to ZATO[b]

Notes:
[a] Other information indicates that it was established in 1994 (Ushakov, 2002, p. 107).
[b] As for ZATO, see Brock, 1998, 2000.

Sources: Ekspert, p. 57, No. 45, 2003 and Ushakov, 2002, pp. 86, 108.

List of Contributors

Masaaki Kuboniwa is Professor at the Institute of Economic Research, Hitotsubashi University, Japan

Yasushi Nakamura is Professor in the Faculty of Economics, Yokohama National University, Japan

Toshihiko Shiobara is Associate Professor in the Faculty of Humanities and Economics, Kochi University, Japan

Shinichiro Tabata is Professor at the Slavic Research Center, Hokkaido University, Japan

Akira Uegaki is Dean of the Department of Economics, Seinan Gakuin University, Japan

Nataliya Ustinova is Deputy Chief of the National Accounts Department, Federal State Statistics Service of Russia (Rosstat), Russia

SLAVIC EURASIAN STUDIES

This series of SLAVIC EURASIAN STUDIES develops interdisciplinary and trans-boundary analyses on the evolving Slavic Eurasian areas. Slavic Eurasia covers the post-communist countries and regions, first of all. However, it is not only a geographical term, but also works as a heuristic concept for better and more realistic interpretations of the changing Eurasian continent under the impacts of globalization. Slavic Eurasia is, according to our understanding, a Mega-area, consisting of Meso-areas. Meso-areas emerge from the post-communist spaces and their formation is in various ways and degrees influenced not only by their internal factors but also by external regional integration such as EU enlargement, Islamic recovery, or East Asian economic growth. Therefore, a Meso-area is not a consolidated spatial framework, but rather a hypothetical term to understand emerging identities in a meso-level between the local or national level and a Mega-area level. Thus Slavic Eurasia, a Mega-area, in turn, loosely binds Meso-areas, sharing the communist experiences, other historical heritages, and politico-economic tasks to be solved in their systemic transformation lasting at least for several decades.

Each volume of the series examines some factors of the evolving Slavic Eurasia, and gives credible interpretations on the dynamic relations among Meso-areas, regional integration, the Mega-area and globalization.

Publication of this series and the related research program entitled "Making a Discipline of Slavic Eurasian Studies," are financed by the 21st Century COE grants of the Ministry of Education, Culture, Sports, Sciences and Technology from 2003 to 2008. These projects involve scholars not only domestically but also internationally, and the organizing engine of the program, the Slavic Research Center, Hokkaido University, serves as a worldwide hub for creating a new approach to Slavic Eurasian Studies.

27th November, 2003

Dr. Osamu Ieda, Program Leader
Professor, Slavic Research Center
Hokkaido University

Slavic Eurasian Studies